Glencoe McGraw-Hill

Homework Practice Workbook

Algebra 2

$f(x) = -0.5x^2$

To the Student

This *Homework Practice Workbook* gives you additional problems for the concept exercises in each lesson. The exercises are designed to aid your study of mathematics by reinforcing important mathematical skills needed to succeed in the everyday world. The materials are organized by chapter and lesson, with one Practice worksheet for every lesson in *Glencoe Algebra 2*.

To the Teacher

These worksheets are the same ones found in the Chapter Resource Masters for *Glencoe Algebra 2*. The answers to these worksheets are available at the end of each Chapter Resource Masters booklet.

The McGraw·Hill Companies

Mc Graw Hill Glencoe

Send all inquiries to:
Glencoe/McGraw-Hill
8787 Orion Place
Columbus, OH 43240-4027

ISBN: 978-0-07-890862-0
MHID: 0-07-890862-0

Homework Practice Workbook, Algebra 2

Printed in the United States of America.

19 20 QTN 16

Contents

Lesson/Title		Page

NAME ______________ DATE ______________ PERIOD ______________

1-1 Skills Practice

Expressions and Formulas

Evaluate each expression if $a = -4$, $b = 6$, and $c = -9$.

1. $3ab - 2bc$

2. $a^3 + c^2 - 3b$

3. $2ac - 12b$

4. $b(a - c) - 2b$

5. $\frac{ac}{b} + \frac{2b}{a}$

6. $\frac{3b - 4c}{2b - (c - b)}$

7. $\frac{3ab}{c} + \frac{2c}{b}$

8. $\frac{b^2}{ac} - c$

Evaluate each expression if $r = -1$, $n = 3$, $t = 12$, $v = 0$, and $w = -\frac{1}{2}$.

9. $6r + 2n$

10. $2nt - 4rn$

11. $w(n - r)$

12. $n + 2r - 16v$

13. $(4n)^2$

14. $n^2r - wt$

15. $2(3r + w)$

16. $\frac{3v + t}{5n - t}$

17. $-w[t + (t - r)]$

18. $\frac{rv^3}{n^2}$

19. $9r^2 + (n^2 - 1)t$

20. $7n - 2v + \frac{2w}{r}$

21. TEMPERATURE The formula $K = C + 273$ gives the temperature in kelvins (K) for a given temperature in degrees Celsius. What is the temperature in kelvins when the temperature is 55 degrees Celsius?

22. TEMPERATURE The formula $C = \frac{5}{9}(F - 32)$ gives the temperature in degrees Celsius for a given temperature in degrees Fahrenheit. What is the temperature in degrees Celsius when the temperature is 68 degrees Fahrenheit?

NAME ______________________ DATE __________ PERIOD __________

1-1 Practice

Expressions and Formulas

Evaluate each expression.

1. $3(4 - 7) - 11$

2. $4(12 - 4^2)$

3. $1 + 2 - 3(4) \div 2$

4. $12 - [20 - 2(6^2 \div 3 \times 2^2)]$

5. $20 \div (5 - 3) + 5^2(3)$

6. $(-2)^3 - (3)(8) + (5)(10)$

7. $18 - \{5 - [34 - (17 - 11)]\}$

8. $[4(5 - 3) - 2(4 - 8)] \div 16$

9. $\frac{1}{2}[6 - 4^2]$

10. $\frac{1}{4}[-5 + 5(-3)]$

11. $\frac{-8(13 - 37)}{6}$

12. $\frac{(-8)^2}{5 - 9} - (-1)^2 + 4(-9)$

Evaluate each expression if $a = \frac{3}{4}$, $b = -8$, $c = -2$, $d = 3$, and $g = \frac{1}{3}$.

13. $ab^2 - d$

14. $(c + d)b$

15. $\frac{ab}{c} + d^2$

16. $\frac{d(b - c)}{ac}$

17. $(b - dg)g^2$

18. $ac^3 - b^2dg$

19. $-b[a + (c - d)^2]$

20. $\frac{ac^4}{d} - \frac{c}{g^2}$

21. $9bc - \frac{1}{g}$

22. $2ab^2 - (d^3 - c)$

23. TEMPERATURE The formula $F = \frac{9}{5}C + 32$ gives the temperature in degrees Fahrenheit for a given temperature in degrees Celsius. What is the temperature in degrees Fahrenheit when the temperature is -40 degrees Celsius?

24. PHYSICS The formula $h = 120t - 16t^2$ gives the height h in feet of an object t seconds after it is shot upward from Earth's surface with an initial velocity of 120 feet per second. What will the height of the object be after 6 seconds?

25. AGRICULTURE Faith owns an organic apple orchard. From her experience the last few seasons, she has developed the formula $P = 20x - 0.01x^2 - 240$ to predict her profit P in dollars this season if her trees produce x bushels of apples. What is Faith's predicted profit this season if her orchard produces 300 bushels of apples?

NAME ______________________ DATE __________ PERIOD __________

1-2 Skills Practice

Properties of Real Numbers

Name the sets of numbers to which each number belongs.

1. 34

2. -525

3. 0.875

4. $\frac{12}{3}$

5. $-\sqrt{9}$

6. $\sqrt{30}$

Name the property illustrated by each equation.

7. $3 \cdot x = x \cdot 3$

8. $3a + 0 = 3a$

9. $2(r + w) = 2r + 2w$

10. $2r + (3r + 4r) = (2r + 3r) + 4r$

11. $5y\left(\frac{1}{5y}\right) = 1$

12. $15x(1) = 15x$

13. $0.6[25(0.5)] = [0.6(25)]0.5$

14. $(10b + 12b) + 7b = (12b + 10b) + 7b$

Find the additive inverse and multiplicative inverse for each number.

15. 15

16. 1.25

17. $-\frac{4}{5}$

18. $3\frac{3}{4}$

Simplify each expression.

19. $3x + 5y + 2x - 3y$

20. $x - y - z + y - x + z$

21. $-(3g + 3h) + 5g - 10h$

22. $a^2 - a + 4a - 3a^2 + 1$

23. $3(m - z) + 5(2m - z)$

24. $2x - 3y - (5x - 3y - 2z)$

25. $6(2w + v) - 4(2v + 1w)$

26. $\frac{1}{3}(15d + 3c) - \frac{1}{2}(8c - 10d)$

NAME ______________________ DATE ____________ PERIOD _____

1-2 Practice

Properties of Real Numbers

Name the sets of numbers to which each number belongs.

1. 6425 **2.** $\sqrt{7}$ **3.** 2π **4.** 0

5. $\sqrt{\frac{25}{36}}$ **6.** $-\sqrt{16}$ **7.** -35 **8.** -31.8

Name the property illustrated by each equation.

9. $5x \cdot (4y + 3x) = 5x \cdot (3x + 4y)$

10. $7x + (9x + 8) = (7x + 9x) + 8$

11. $5(3x + y) = 5(3x + 1y)$

12. $7n + 2n = (7 + 2)n$

13. $3(2x)y = (3 \cdot 2)(xy)$

14. $3x \cdot 2y = 3 \cdot 2 \cdot x \cdot y$

15. $(6 + -6)y = 0y$

16. $\frac{1}{4} \cdot 4y = 1y$

17. $5(x + y) = 5x + 5y$

18. $4n + 0 = 4n$

Find the additive inverse and multiplicative inverse for each number.

19. 0.4

20. -1.6

21. $-\frac{11}{16}$

22. $5\frac{5}{6}$

Simplify each expression.

23. $5x - 3y - 2x + 3y$

24. $-11a - 13b + 7a - 3b$

25. $8x - 7y - (3 - 6y)$

26. $4c - 2c - (4c + 2c)$

27. $3(r - 10t) - 4(7t + 2r)$

28. $\frac{1}{5}(10a - 15b) + \frac{1}{2}(8b + 4a)$

29. $2(4z - 2x + y) - 4(5z + x - y)$

30. $\frac{5}{6}\left(\frac{3}{5}x + 12y\right) - \frac{1}{4}(2x - 12y)$

31. TRAVEL Olivia drives her car at 60 miles per hour for t hours. Ian drives his car at 50 miles per hour for $(t + 2)$ hours. Write a simplified expression for the sum of the distances traveled by the two cars.

32. NUMBER THEORY Use the properties of real numbers to tell whether the following statement is true or false: If a and $b \neq 0$ and $a > b$, it follows that $a\left(\frac{1}{a}\right) > b\left(\frac{1}{b}\right)$. Explain your reasoning.

NAME ______________________ DATE __________ PERIOD __________

1-3 Skills Practice

Solving Equations

Write an algebraic expression to represent each verbal expression.

1. 4 times a number, increased by 7

2. 8 less than 5 times a number

3. 6 times the sum of a number and 5

4. the product of 3 and a number, divided by 9

5. 3 times the difference of 4 and a number

6. the product of −11 and the square of a number

Write a verbal sentence to represent each equation.

7. $n - 8 = 16$

8. $8 + 3x = 5$

9. $b + 3 = b^2$

10. $\frac{y}{3} = 2 - 2y$

Name the property illustrated by each statement.

11. If $a = 0.5b$, and $0.5b = 10$, then $a = 10$.

12. If $d + 1 = f$, then $d = f - 1$.

13. If $-7x = 14$, then $14 = -7x$.

14. If $(8 + 7)r = 30$, then $15r = 30$.

Solve each equation. Check your solution.

15. $4m + 2 = 18$

16. $x + 4 = 5x + 2$

17. $3t = 2t + 5$

18. $-3b + 7 = -15 + 2b$

19. $-5x = 3x - 24$

20. $4v + 20 - 6 = 34$

21. $a - \frac{2a}{5} = 3$

22. $2.2n + 0.8n + 5 = 4n$

Solve each equation or formula for the specified variable.

23. $I = prt$, for p

24. $y = \frac{1}{4}x - 12$, for x

25. $A = \frac{x + y}{2}$, for y

26. $A = 2\pi r^2 + 2\pi rh$, for h

NAME ______________________ DATE __________ PERIOD ________

1-3 Practice

Solving Equations

Write an algebraic expression to represent each verbal expression.

1. 2 more than the quotient of a number and 5

2. the sum of two consecutive integers

3. 5 times the sum of a number and 1

4. 1 less than twice the square of a number

Write a verbal sentence to represent each equation.

5. $5 - 2x = 4$

6. $3y = 4y^3$

7. $3c = 2(c - 1)$

8. $\frac{m}{5} = 3(2m + 1)$

Name the property illustrated by each statement.

9. If $t - 13 = 52$, then $52 = t - 13$.

10. If $8(2q + 1) = 4$, then $2(2q + 1) = 1$.

11. If $h + 12 = 22$, then $h = 10$.

12. If $4m = -15$, then $-12m = 45$.

Solve each equation. Check your solution.

13. $14 = 8 - 6r$

14. $9 + 4n = -59$

15. $\frac{3}{4} - \frac{1}{2}n = \frac{5}{8}$

16. $\frac{5}{6}c + \frac{3}{4} = \frac{11}{12}$

17. $-1.6r + 5 = -7.8$

18. $6x - 5 = 7 - 9x$

19. $5(6 - 4v) = v + 21$

20. $6y - 5 = -3(2y + 1)$

Solve each equation or formula for the specified variable.

21. $E = mc^2$, for m

22. $c = \frac{2d + 1}{3}$, for d

23. $h = vt - gt^2$, for v

24. $E = \frac{1}{2}Iw^2 + U$, for I

25. GEOMETRY The length of a rectangle is twice the width. Find the width if the perimeter is 60 centimeters. Define a variable, write an equation, and solve the problem.

26. GOLF Luis and three friends went golfing. Two of the friends rented clubs for $6 each. The total cost of the rented clubs and the green fees for each person was $76. What was the cost of the green fees for each person? Define a variable, write an equation, and solve the problem.

NAME ______________________ DATE ____________ PERIOD ____________

1-4 Skills Practice

Solving Absolute Value Equations

Evaluate each expression if $w = 0.4$, $x = 2$, $y = -3$, and $z = -10$.

1. $|5w|$

2. $|-9y|$

3. $|9y - z|$

4. $-|17z|$

5. $-|10z - 31|$

6. $-|8x - 3y| + |2y + 5x|$

7. $25 - |5z + 1|$

8. $44 + |-2x - y|$

9. $2|4w|$

10. $3 - |1 - 6w|$

11. $|-3x - 2y| - 4$

12. $6.4 + |w - 1|$

Solve each equation. Check your solutions.

13. $|y + 3| = 2$

14. $|5a| = 10$

15. $|3k - 6| = 2$

16. $|2g + 6| = 0$

17. $10 = |1 - c|$

18. $|2x + x| = 9$

19. $|p - 7| = -14$

20. $2|3w| = 12$

21. $|7x - 3x| + 2 = 18$

22. $4|7 - y| - 1 = 11$

23. $|3n - 2| = \frac{1}{2}$

24. $|8d - 4d| + 5 = 13$

25. $-5|6a + 2| = -15$

26. $|k| + 10 = 9$

NAME ______________________ DATE ____________ PERIOD ________

1-4 Practice

Solving Absolute Value Equations

Evaluate each expression if $a = -1$, $b = -8$, $c = 5$, and $d = -1.4$.

1. $|6a|$

2. $|2b + 4|$

3. $-|10d + a|$

4. $|17c| + |3b - 5|$

5. $-6|10a - 12|$

6. $|2b - 1| - |-8b + 5|$

7. $|5a - 7| + |3c - 4|$

8. $|1 - 7c| - |a|$

9. $-3|0.5c + 2| - |-0.5b|$

10. $|4d| + |5 - 2a|$

11. $|a - b| + |b - a|$

12. $|2 - 2d| - 3|b|$

Solve each equation. Check your solutions.

13. $|n - 4| = 13$

14. $|x - 13| = 2$

15. $|2y - 3| = 29$

16. $7|x + 3| = 42$

17. $|3u - 6| = 42$

18. $|5x - 4| = -6$

19. $-3|4x - 9| = 24$

20. $-6|5 - 2y| = -9$

21. $|8 + p| = 2p - 3$

22. $|4w - 1| = 5w + 37$

23. $4|2y - 7| + 5 = 9$

24. $-2|7 - 3y| - 6 = -14$

25. $2|4 - m| = -3m$

26. $5 - 3|2 + 2w| = -7$

27. $5|2r + 3| - 5 = 0$

28. $3 - 5|2d - 3| = 4$

29. WEATHER A thermometer comes with a guarantee that the stated temperature differs from the actual temperature by no more than 1.5 degrees Fahrenheit. Write and solve an equation to find the minimum and maximum actual temperatures when the thermometer states that the temperature is 87.4 degrees Fahrenheit.

30. OPINION POLLS Public opinion polls reported in newspapers are usually given with a margin of error. For example, a poll with a margin of error of ±5% is considered accurate to within plus or minus 5% of the actual value. A poll with a stated margin of error of 63% predicts that candidate Tonwe will receive 51% of an upcoming vote. Write and solve an equation describing the minimum and maximum percent of the vote that candidate Tonwe is expected to receive.

NAME ______________ DATE ______ PERIOD ______

1-5 Skills Practice

Solving Inequalities

Solve each inequality. Then graph the solution set on a number line.

1. $\frac{z}{-4} \geq 2$

−9 −8 −7 −6 −5 −4 −3 −2 −1

2. $3a + 7 \leq 16$

−4 −3 −2 −1 0 1 2 3 4

3. $16 < 3q + 4$

−1 0 1 2 3 4 5 6 7

4. $20 - 3n > 7n$

−4 −3 −2 −1 0 1 2 3 4

5. $3x \geq -9$

−4 −3 −2 −1 0 1 2 3 4

6. $4b - 9 \leq 7$

−2 −1 0 1 2 3 4 5 6

7. $2z < -9 + 5z$

−2 −1 0 1 2 3 4 5 6

8. $7f - 9 > 3f - 1$

−4 −3 −2 −1 0 1 2 3 4

9. $-3k - 8 \leq 5k$

−4 −3 −2 −1 0 1 2 3 4

10. $7t - (t - 4) \leq 25$

−4 −3 −2 −1 0 1 2 3 4

11. $0.7m + 0.3m \geq 2m - 4$

−2 −1 0 1 2 3 4 5 6

12. $4(5x + 7) \leq 13$

−4 −3 −2 −1 0 1 2 3 4

13. $1.7y - 0.78 > 5$

−2 −1 0 1 2 3 4 5 6

14. $4x - 9 > 2x + 1$

−1 0 1 2 3 4 5 6 7

Define a variable and write an inequality for each problem. Then solve.

15. Nineteen more than a number is less than 42.

16. The difference of three times a number and 16 is at least 8.

17. One half of a number is more than 6 less than the same number.

18. Five less than the product of 6 and a number is no more than twice that same number.

NAME ______________________________ DATE ______________ PERIOD ____________

1-5 Practice

Solving Inequalities

Solve each inequality. Then graph the solution set on a number line.

1. $8x - 6 \geq 10$

−4 −3 −2 −1 0 1 2 3 4

2. $23 - 4u < 11$

−2 −1 0 1 2 3 4 5 6

3. $-16 - 8r \geq 0$

−4 −3 −2 −1 0 1 2 3 4

4. $14c < 9c + 5$

−4 −3 −2 −1 0 1 2 3 4

5. $9x - 11 > 6x - 9$

−4 −3 −2 −1 0 1 2 3 4

6. $-3(4w - 1) > 18$

−4 −3 −2 −1 0 1 2 3 4

7. $1 - 8u \leq 3u - 10$

−4 −3 −2 −1 0 1 2 3 4

8. $17.5 < 19 - 2.5x$

−4 −3 −2 −1 0 1 2 3 4

9. $9(2r - 5) - 3 < 7r - 4$

10. $1 + 5(x - 8) \leq 2 - (x + 5)$

11. $\frac{4x - 3}{2} \geq -3.5$

12. $q - 2(2 - q) \leq 0$

13. $-36 - 2(w + 77) > -4(2w + 52)$

14. $4n - 5(n - 3) > 3(n + 1) - 4$

Define a variable and write an inequality for each problem. Then solve.

15. Twenty less than a number is more than twice the same number.

16. Four times the sum of twice a number and −3 is less than 5.5 times that same number.

17. HOTELS The Lincoln's hotel room costs $90 a night. An additional 10% tax is added. Hotel parking is $12 per day. The Lincoln's expect to spend $30 in tips during their stay. Solve the inequality $90x + 90(0.1)x + 12x + 30 \leq 600$ to find how many nights the Lincoln's can stay at the hotel without exceeding total hotel costs of $600.

18. BANKING Jan's account balance is $3800. Of this, $750 is for rent. Jan wants to keep a balance of at least $500. Write and solve an inequality describing how much she can withdraw and still leave enough for rent and a $500 balance.

NAME ______________________ DATE __________ PERIOD __________

1-6 Skills Practice

Solving Compound and Absolute Value Inequalities

Write an absolute value inequality for each graph.

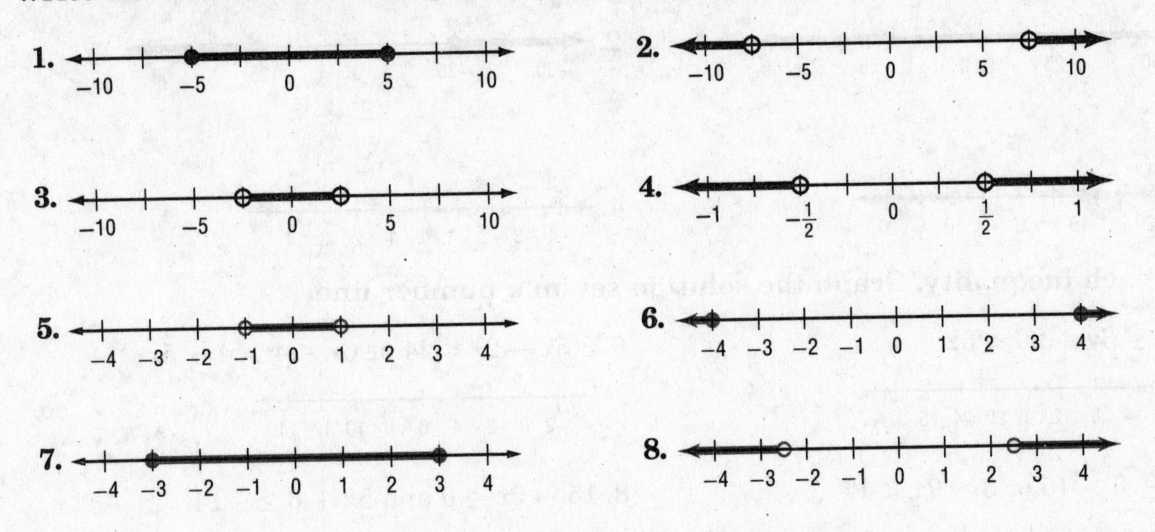

Solve each inequality. Graph the solution set on a number line.

9. $2c + 1 > 5$ or $c < 0$

–4 –3 –2 –1 0 1 2 3 4

10. $-11 \leq 4y - 3 \leq 1$

–4 –3 –2 –1 0 1 2 3 4

11. $10 > -5x > 5$

–4 –3 –2 –1 0 1 2 3 4

12. $4a \geq -8$ or $a < -3$

–4 –3 –2 –1 0 1 2 3 4

13. $8 < 3x + 2 \leq 23$

0 1 2 3 4 5 6 7 8

14. $w - 4 \leq 10$ or $-2w \leq 6$

–5 0 5 10 15

15. $|t| \geq 3$

–4 –3 –2 –1 0 1 2 3 4

16. $|6x| < 12$

–4 –3 –2 –1 0 1 2 3 4

17. $|-7r| > 14$

–4 –3 –2 –1 0 1 2 3 4

18. $|p + 2| \leq -2$

–4 –3 –2 –1 0 1 2 3 4

19. $|n - 5| < 7$

–4 –2 0 2 4 6 8 10 12

20. $|h + 1| \geq 5$

–8 –6 –4 –2 0 2 4 6 8

NAME ______________________ DATE __________ PERIOD __________

1-6 Practice

Solving Compound and Absolute Value Inequalities

Write an absolute value inequality for each graph.

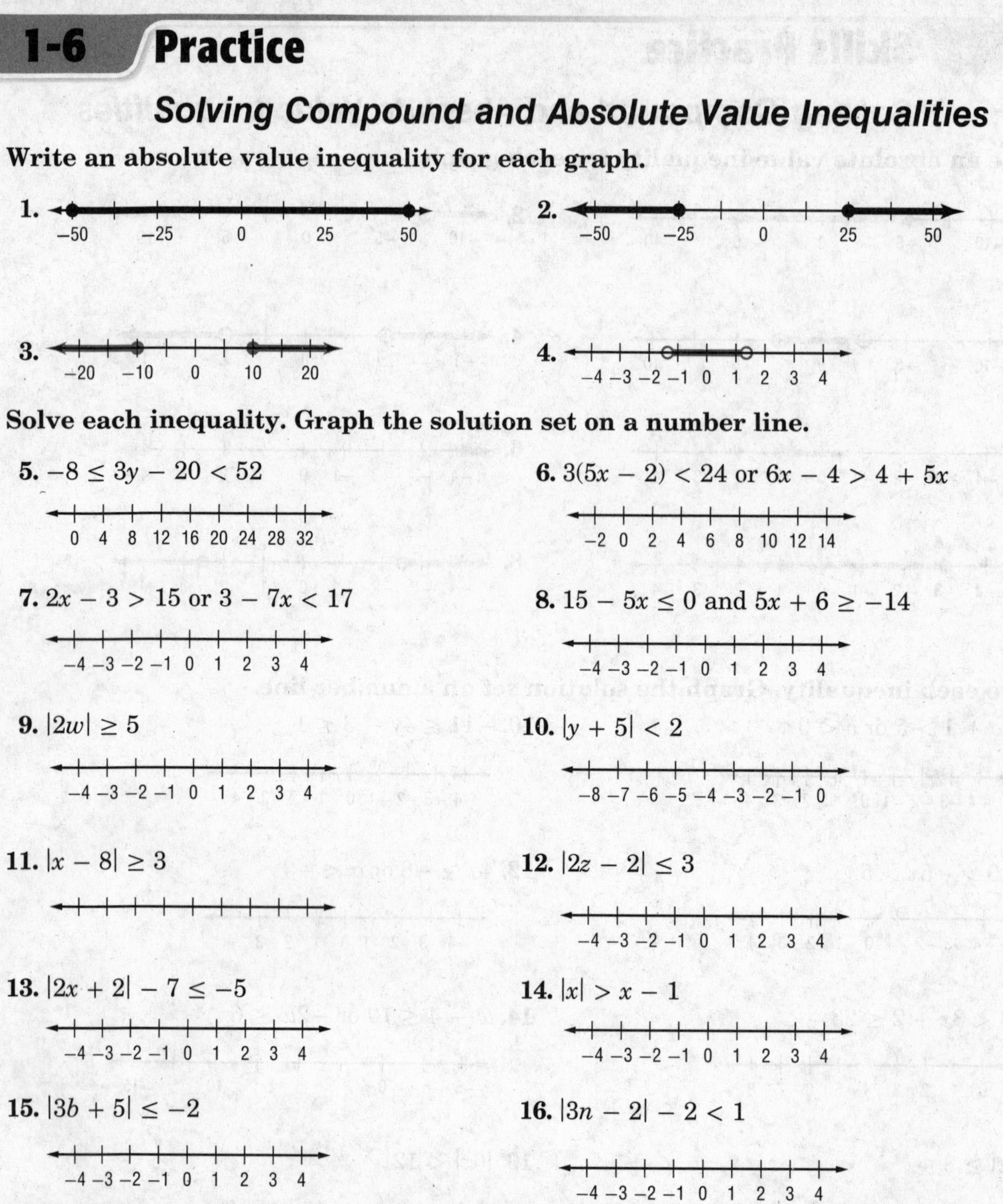

Solve each inequality. Graph the solution set on a number line.

5. $-8 \leq 3y - 20 < 52$

0 4 8 12 16 20 24 28 32

6. $3(5x - 2) < 24$ or $6x - 4 > 4 + 5x$

−2 0 2 4 6 8 10 12 14

7. $2x - 3 > 15$ or $3 - 7x < 17$

−4 −3 −2 −1 0 1 2 3 4

8. $15 - 5x \leq 0$ and $5x + 6 \geq -14$

−4 −3 −2 −1 0 1 2 3 4

9. $|2w| \geq 5$

−4 −3 −2 −1 0 1 2 3 4

10. $|y + 5| < 2$

−8 −7 −6 −5 −4 −3 −2 −1 0

11. $|x - 8| \geq 3$

12. $|2z - 2| \leq 3$

−4 −3 −2 −1 0 1 2 3 4

13. $|2x + 2| - 7 \leq -5$

−4 −3 −2 −1 0 1 2 3 4

14. $|x| > x - 1$

−4 −3 −2 −1 0 1 2 3 4

15. $|3b + 5| \leq -2$

−4 −3 −2 −1 0 1 2 3 4

16. $|3n - 2| - 2 < 1$

−4 −3 −2 −1 0 1 2 3 4

17. RAINFALL In 90% of the last 30 years, the rainfall at Shell Beach has varied no more than 6.5 inches from its mean value of 24 inches. Write and solve an absolute value inequality to describe the rainfall in the other 10% of the last 30 years.

18. MANUFACTURING A company's guidelines call for each can of soup produced not to vary from its stated volume of 14.5 fluid ounces by more than 0.08 ounces. Write and solve an absolute value inequality to describe acceptable can volumes.

NAME ______________________ DATE __________ PERIOD __________

2-1 Skills Practice

Relations and Functions

State the domain and range of each relation. Then determine whether each relation is a *function*. If it is a function, determine if it is *one-to-one, onto, both* or *neither*.

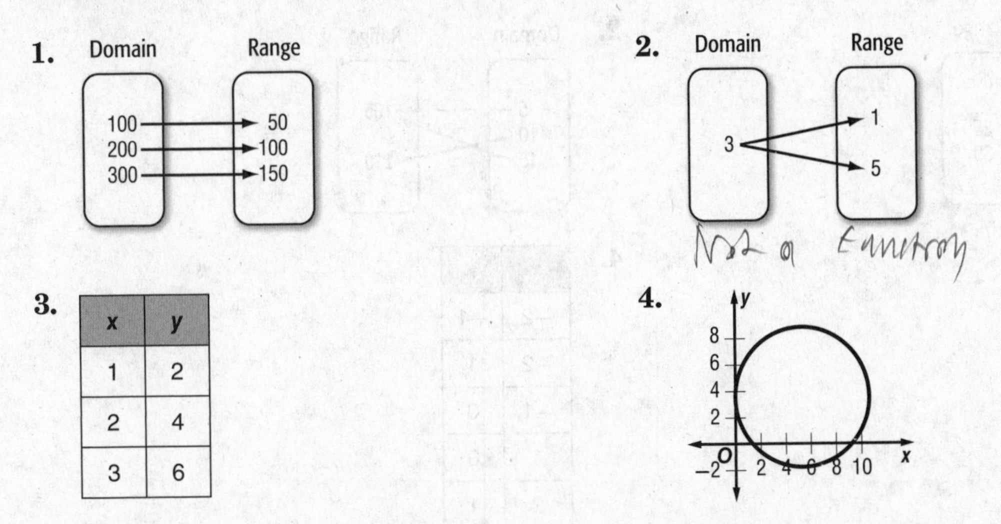

x	y
1	2
2	4
3	6

Graph each relation or equation and determine the domain and range. Determine whether the eqation is a *function*, is *one-to-one, onto, both*, or *neither*. Then state whether it is discrete or continuous.

5. $\{(2, -3), (2, 4), (2, -1)\}$

6. $\{(2, 6), (6, 2)\}$

7. $\{(-3, 4), (-2, 4), (-1, -1), (3, -1)\}$

8. $x = -2$

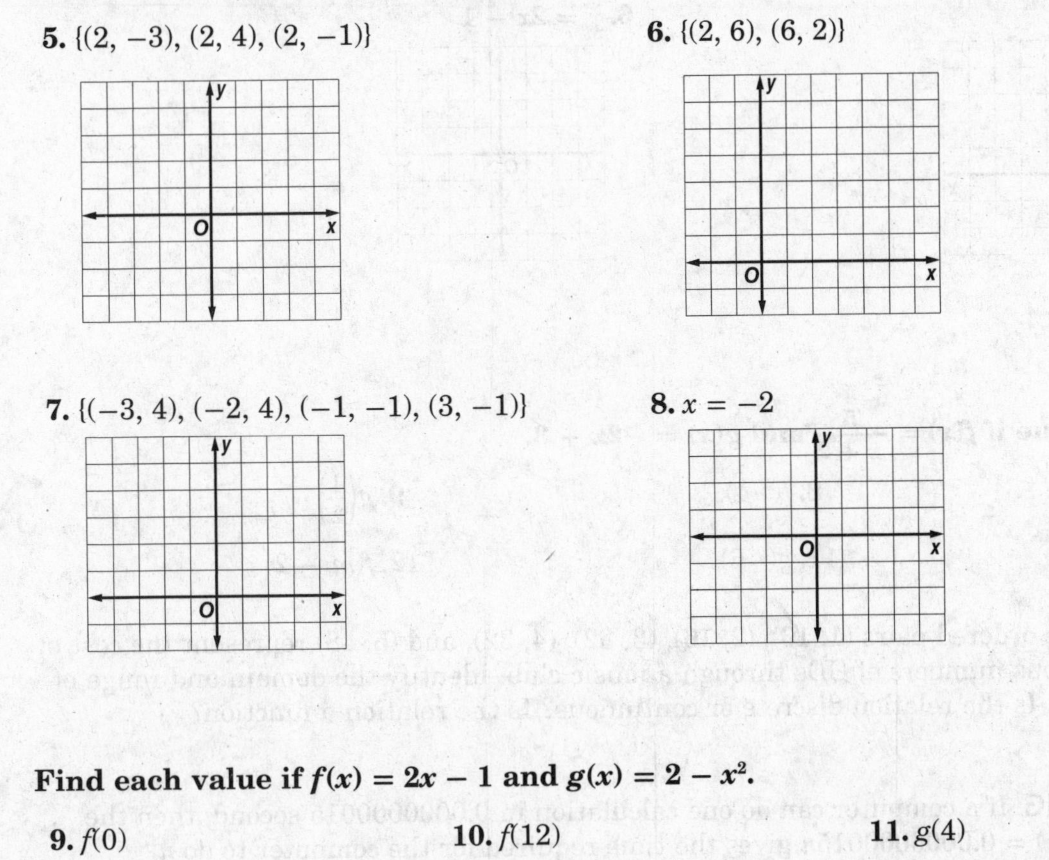

Find each value if $f(x) = 2x - 1$ and $g(x) = 2 - x^2$.

9. $f(0)$

10. $f(12)$

11. $g(4)$

12. $f(-2)$

13. $g(-1)$

14. $f(d)$

2-1 Practice

Relations and Functions

State the domain and range of each relation. Then determine whether each relation is a *function*. If it is a function, determine if it is *one-to-one, onto, both* or *neither*.

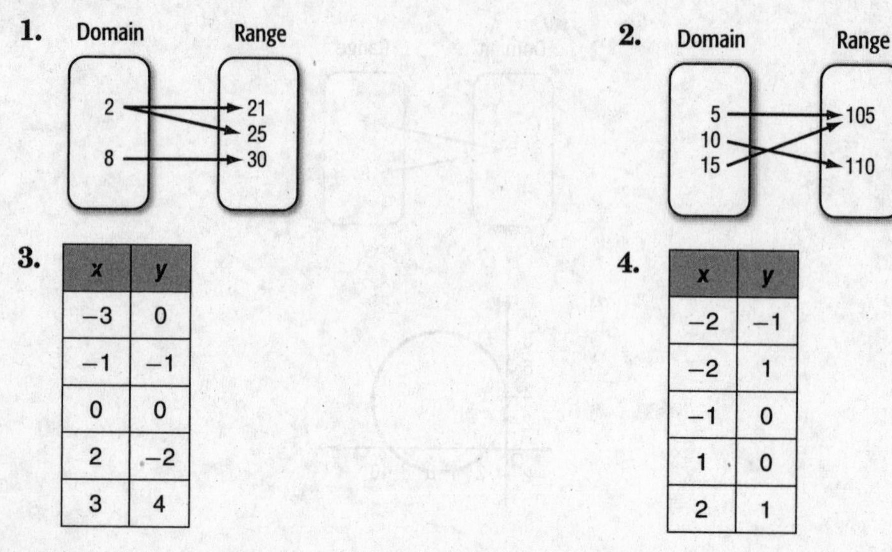

3.

x	y
−3	0
−1	−1
0	0
2	−2
3	4

4.

x	y
−2	−1
−2	1
−1	0
1	0
2	1

Graph each equation and determine the domain and range. Determine whether the relation is a *function*, is *one-to-one, onto, both*, or *neither*. Then state whether it is *discrete* or *continuous*.

5. $x = -1$

6. $y = 2x - 1$

Find each value if $f(x) = \frac{5}{x+2}$ and $g(x) = -2x + 3$.

7. $f(3)$

8. $f(-4)$

9. $g\left(\frac{1}{2}\right)$

10. $f(-2)$

11. $g(-6)$

12. $f(m - 2)$

13. **MUSIC** The ordered pairs (1, 16), (2, 16), (3, 32), (4, 32), and (5, 48) represent the cost of buying various numbers of CDs through a music club. Identify the domain and range of the relation. Is the relation discrete or continuous? Is the relation a function?

14. **COMPUTING** If a computer can do one calculation in 0.0000000015 second, then the function $T(n) = 0.0000000015n$ gives the time required for the computer to do n calculations. How long would it take the computer to do 5 billion calculations?

2-2 Skills Practice

Linear Relations and Functions

State whether each function is a linear function. Explain.

1. $y = 3x$

2. $y = -2 + 5x$

3. $2x + y = 10$

4. $f(x) = 4x^2$

5. $-\frac{3}{x} + y = 15$

6. $x = y + 8$

7. $g(x) = 8$

8. $h(x) = \sqrt{x} + 3$

Write each equation in standard form. Identify *A*, *B*, and *C*.

9. $y = x$

10. $y = 5x + 1$

11. $2x = 4 - 7y$

12. $3x = -2y - 2$

13. $5y - 9 = 0$

14. $-6y + 14 = 8x$

Find the *x*-intercept and the *y*-intercept of the graph of each equation. Then graph the equation using the intercepts.

15. $y = 3x - 6$

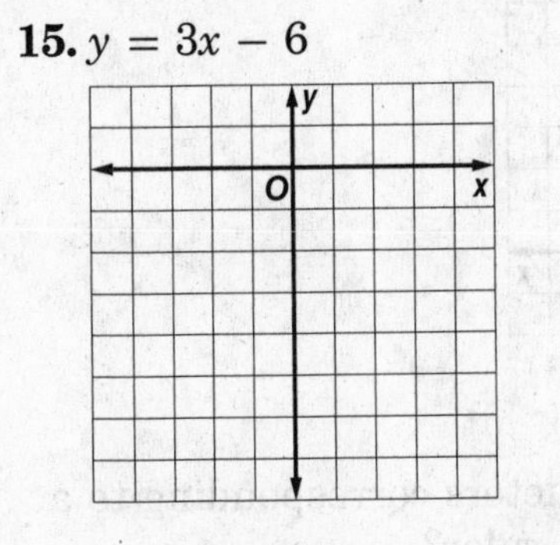

16. $y = -2x$

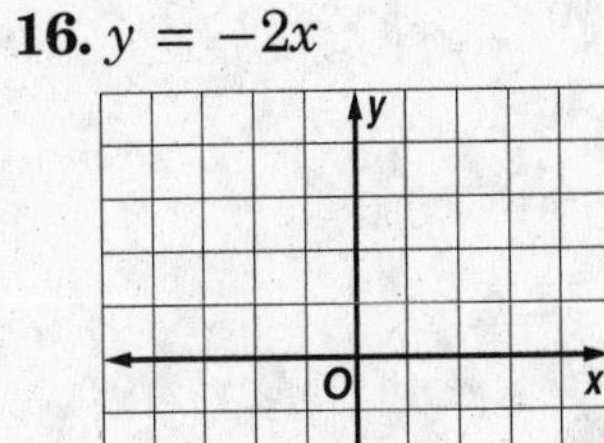

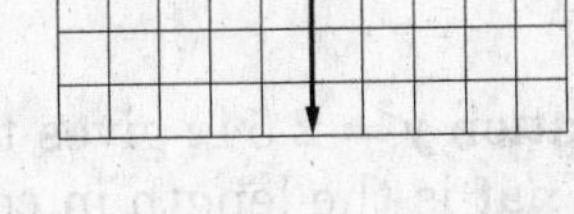

17. $x + y = 5$

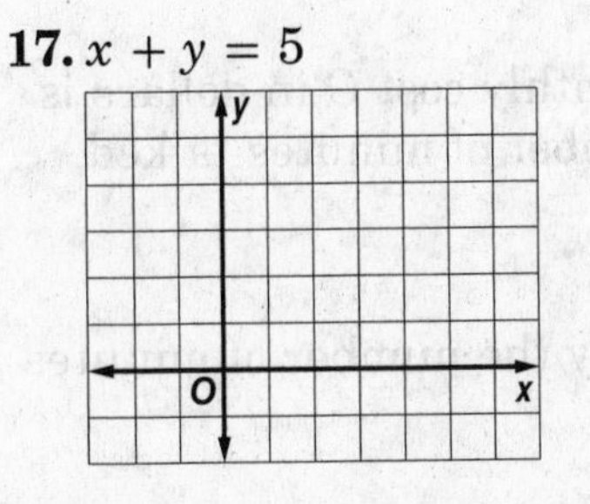

18. $2x + 5y = 10$

NAME ______________________ DATE ____________ PERIOD ________

2-2 Practice

Linear Relations and Functions

State whether each function is a linear function. Write *yes* or *no*. Explain.

1. $h(x) = 23$

2. $y = \frac{2}{3}x$

3. $y = \frac{5}{x}$

4. $9 - 5xy = 2$

Write each equation in standard form. Identify *A*, *B*, and *C*.

5. $y = 7x - 5$

6. $y = \frac{3}{8}x + 5$

7. $3y - 5 = 0$

8. $x = -\frac{2}{7}y + \frac{3}{4}$

Find the *x*-intercept and the *y*-intercept of the graph of each equation. Then graph the equation using the intercepts.

9. $y = 2x + 4$

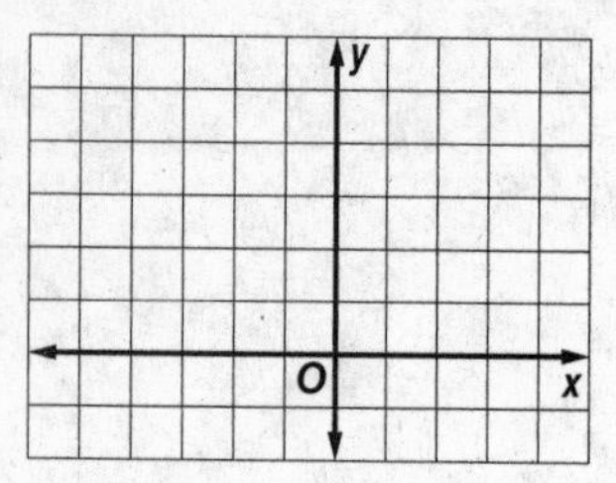

10. $2x + 7y = 14$

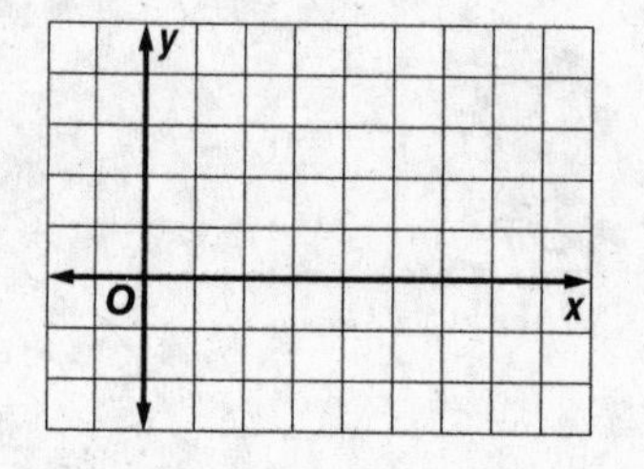

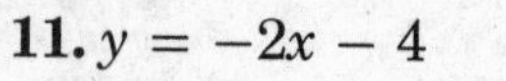

11. $y = -2x - 4$

12. $6x + 2y = 6$

13. MEASURE The equation $y = 2.54x$ gives the length y in centimeters corresponding to a length x in inches. What is the length in centimeters of a 1-foot ruler?

14. LONG DISTANCE For Meg's long-distance calling plan, the monthly cost C in dollars is given by the linear function $C(t) = 6 + 0.05t$, where t is the number of minutes talked.

a. What is the total cost of talking 8 hours? of talking 20 hours?

b. What is the effective cost per minute (the total cost divided by the number of minutes talked) of talking 8 hours? of talking 20 hours?

NAME ______________________ DATE __________ PERIOD __________

2-3 Skills Practice

Rate of Change and Slope

Find the slope of the line that passes through each pair of points.

1. (1, 5), (–1, –3)

2. (0, 2), (3, 0)

3. (1, 9), (0, 6)

4. (8, –5), (4, –2)

5. (–3, 5), (–3, –1)

6. (–2, –2), (10, –2)

7. (4, 5), (2, 7)

8. (–2, –4), (3, 2)

9. (5, 2), (–3, 2)

Determine the rate of change of each graph.

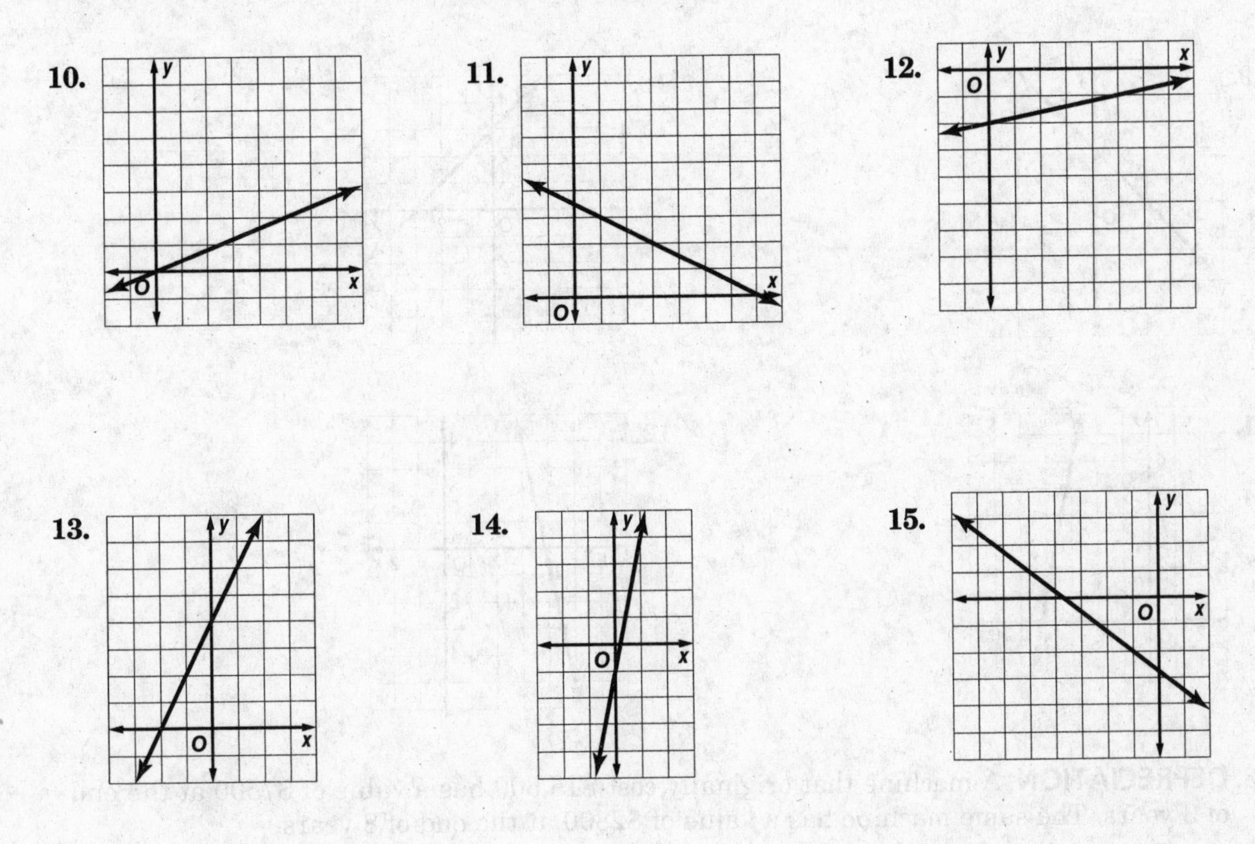

16. HIKING Naomi left from an elevation of 7400 feet at 7:00 A.M. and hiked to an elevation of 9800 feet by 11:00 A.M. What was her rate of change in altitude?

NAME ______________________ DATE ____________ PERIOD ________

2-3 Practice

Rate of Change and Slope

Find the slope of the line that passes through each pair of points. Express as a fraction in simplest form.

1. (3, −8), (−5, 2)

2. (−10, −3), (7, 2)

3. (−7, −6), (3, −6)

4. (8, 2), (8, −1)

5. (4, 3), (7, −2)

6. (−6, −3), (−8, 4)

Determine the rate of change of each graph.

7.

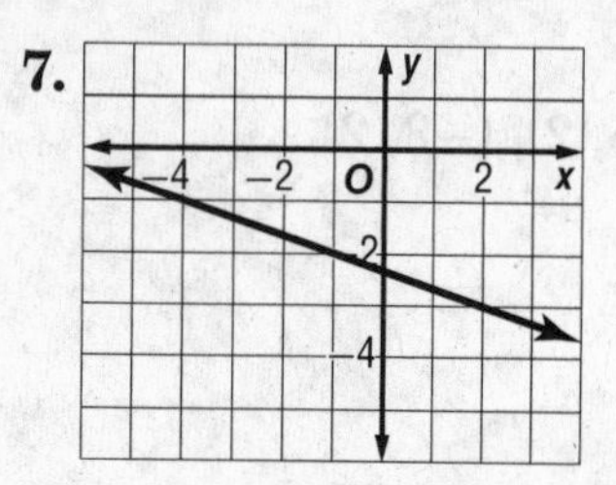

8.

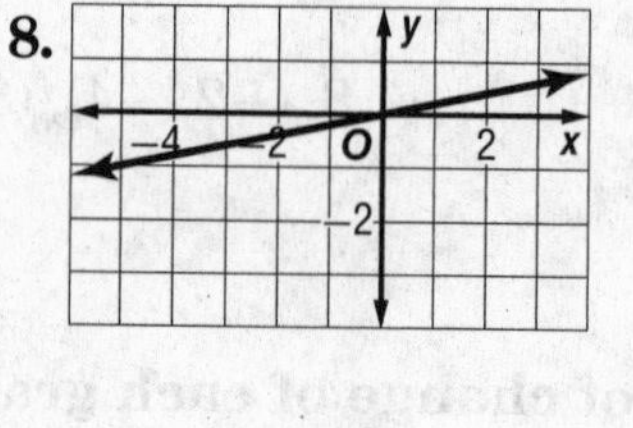

9.

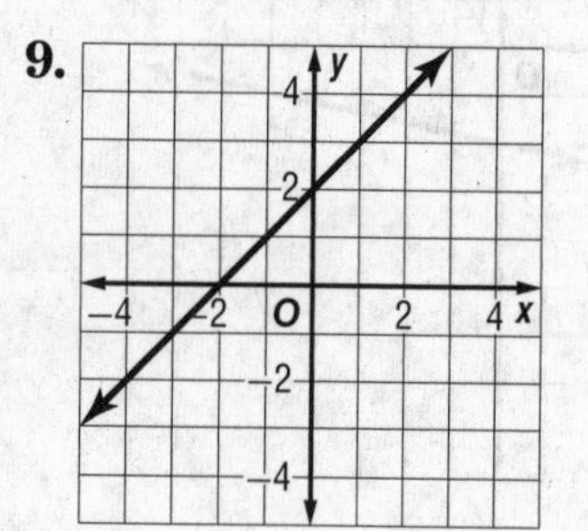

10.

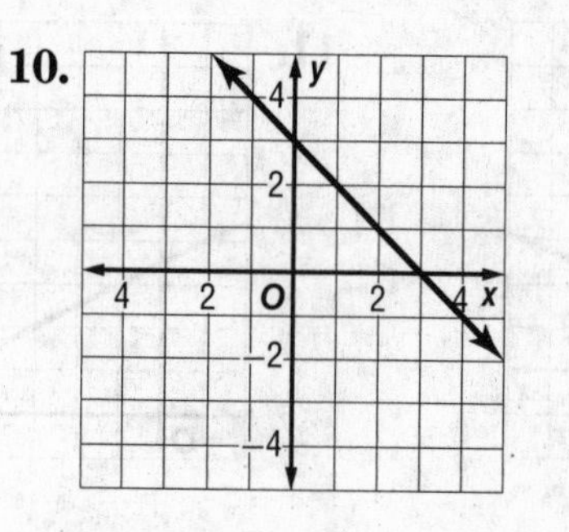

11.

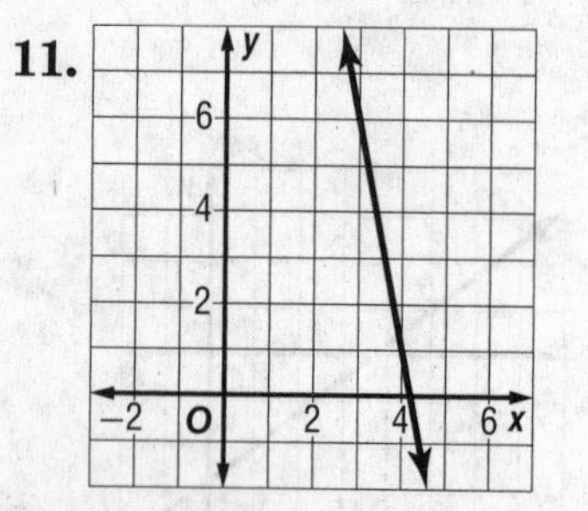

12.

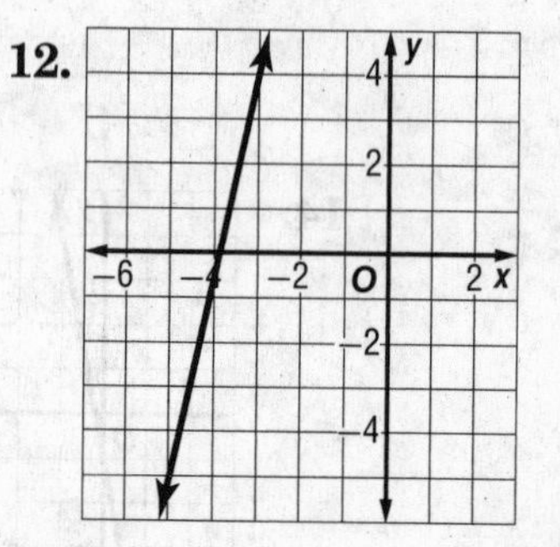

13. DEPRECIATION A machine that originally cost $15,600 has a value of $7500 at the end of 3 years. The same machine has a value of $2800 at the end of 8 years.

a. Find the average rate of change in value (depreciation) of the machine between its purchase and the end of 3 years.

b. Find the average rate of change in value of the machine between the end of 3 years and the end of 8 years.

c. Interpret the sign of your answers.

NAME ______________________ DATE ____________ PERIOD ____________

2-4 Skills Practice

Writing Linear Equations

Write an equation in slope-intercept form for the line described.

1. slope 3, y-intercept at -4

2. perpendicular to $y = \frac{1}{2}x - 1$, x-intercept at 4

3. parallel to $y = \frac{2}{3}x + 6$, passes through (6, 7)

4. parallel to $y = -\frac{1}{4}x - 2$, x-intercept at 4

5. perpendicular to $y = -4x + 1$, passes through $(-8, -1)$

6. slope $\frac{3}{5}$, x-intercept at -10

7. parallel to $y = 9x + 3$, y-intercept at -2

8. slope $\frac{5}{6}$, passes through (12, 4)

Write an equation in slope-intercept form for each graph.

9.

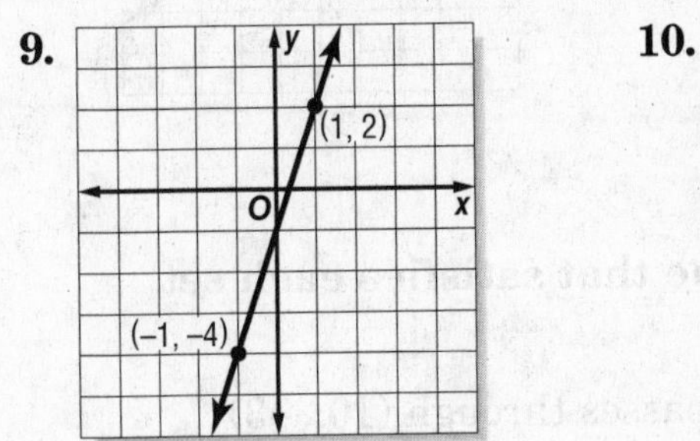

10.

y
O
x
(–3, –1)
(4, –1)

11.

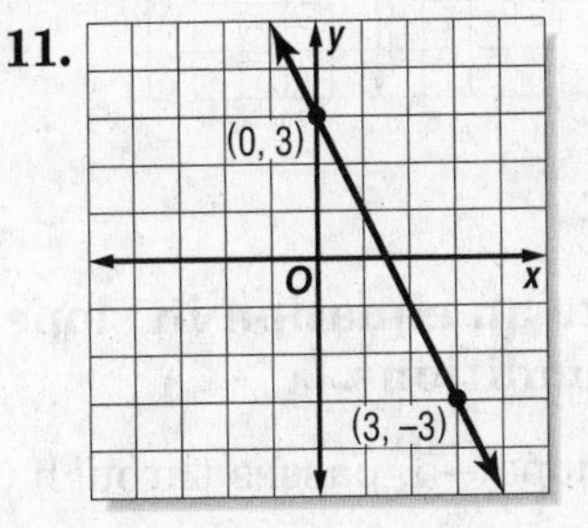

Write an equation in slope-intercept form for the line that satisfies each set of conditions.

12. slope 3, passes through $(1, -3)$

13. slope -1, passes through (0, 0)

14. slope -2, passes through $(0, -5)$

15. slope 3, passes through (2, 0)

16. passes through $(-1, -2)$ and $(-3, 1)$

17. passes through $(-2, -4)$ and (1, 8)

18. passes through (2, 0) and $(0, -6)$

19. passes through (2.5, 0) and (0, 5)

20. passes through $(3, -1)$, perpendicular to the graph of $y = -\frac{1}{3}x - 4$.

NAME ______________________ DATE ____________ PERIOD ____________

2-4 Practice

Writing Linear Equations

Write an equation in slope-intercept form for the line described.

1. slope 2, y-intercept at 0

2. parallel to $y = 4x + 2$, y-intercept at 4

3. perpendicular to $y = \frac{1}{4}x + 2$, passes through (0, 0)

4. parallel to $y = -3x + 4$, x-intercept at 4

5. perpendicular to $y = -\frac{1}{2}x + \frac{2}{3}$, passes through (2, 3)

6. slope $-\frac{2}{3}$, x-intercept at 3

Write an equation in slope-intercept form for each graph.

7.

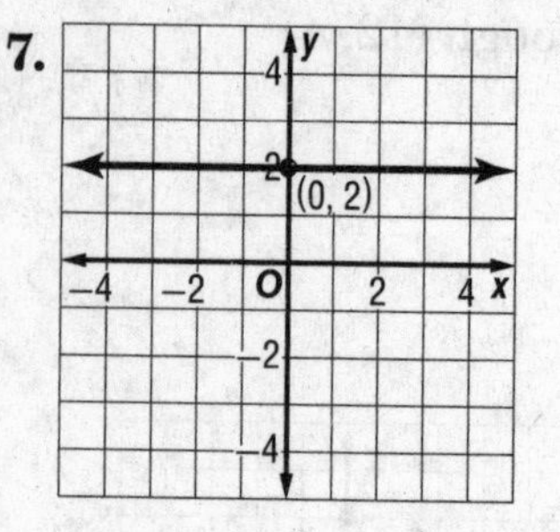

8.

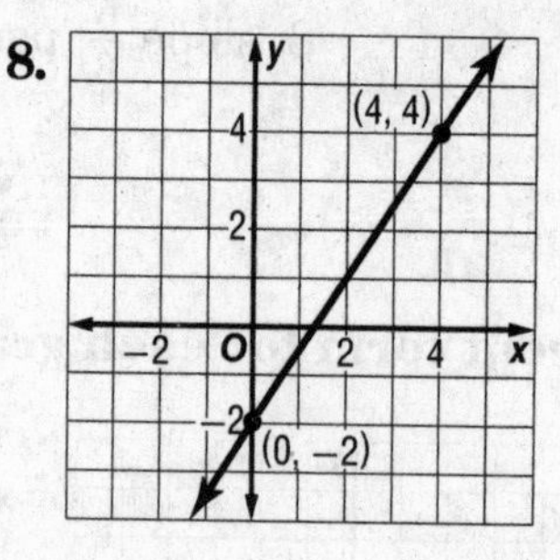

9.

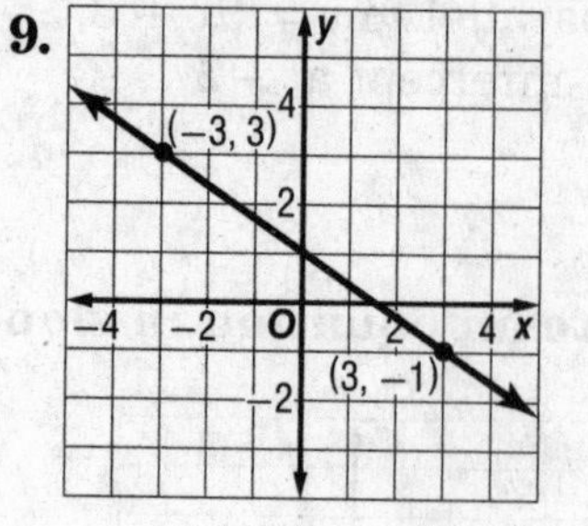

Write an equation in slope-intercept form for the line that satisfies each set of conditions.

10. slope −5, passes through (−3, −8)

11. slope $\frac{4}{5}$, passes through (10, −3)

12. slope 0, passes through (0, −10)

13. slope $-\frac{2}{3}$, passes through (6, −8)

14. parallel to $y = 4x - 5$, y-intercept at −6

15. slope $\frac{1}{6}$, x-intercept at −1

16. perpendicular to $y = 3x - 2$ passes through (6, −1)

17. parallel to $y = \frac{2}{3}x - 10$, x-intercept at 9

18. passes through (−8, −7), perpendicular to the graph of $y = 4x - 3$

19. RESERVOIRS The surface of Grand Lake is at an elevation of 648 feet. During the current drought, the water level is dropping at a rate of 3 inches per day. If this trend continues, write an equation that gives the elevation in feet of the surface of Grand Lake after x days.

NAME _______________ DATE ________ PERIOD ________

2-5 Skills Practice

Scatter Plots and Lines of Regression

For Exercises 1–3, complete parts a–c.

a. Make a scatter plot and a line of fit, and describe the correlation.
b. Use two ordered pairs to write a prediction equation.
c. Use your prediction equation to predict the missing value.

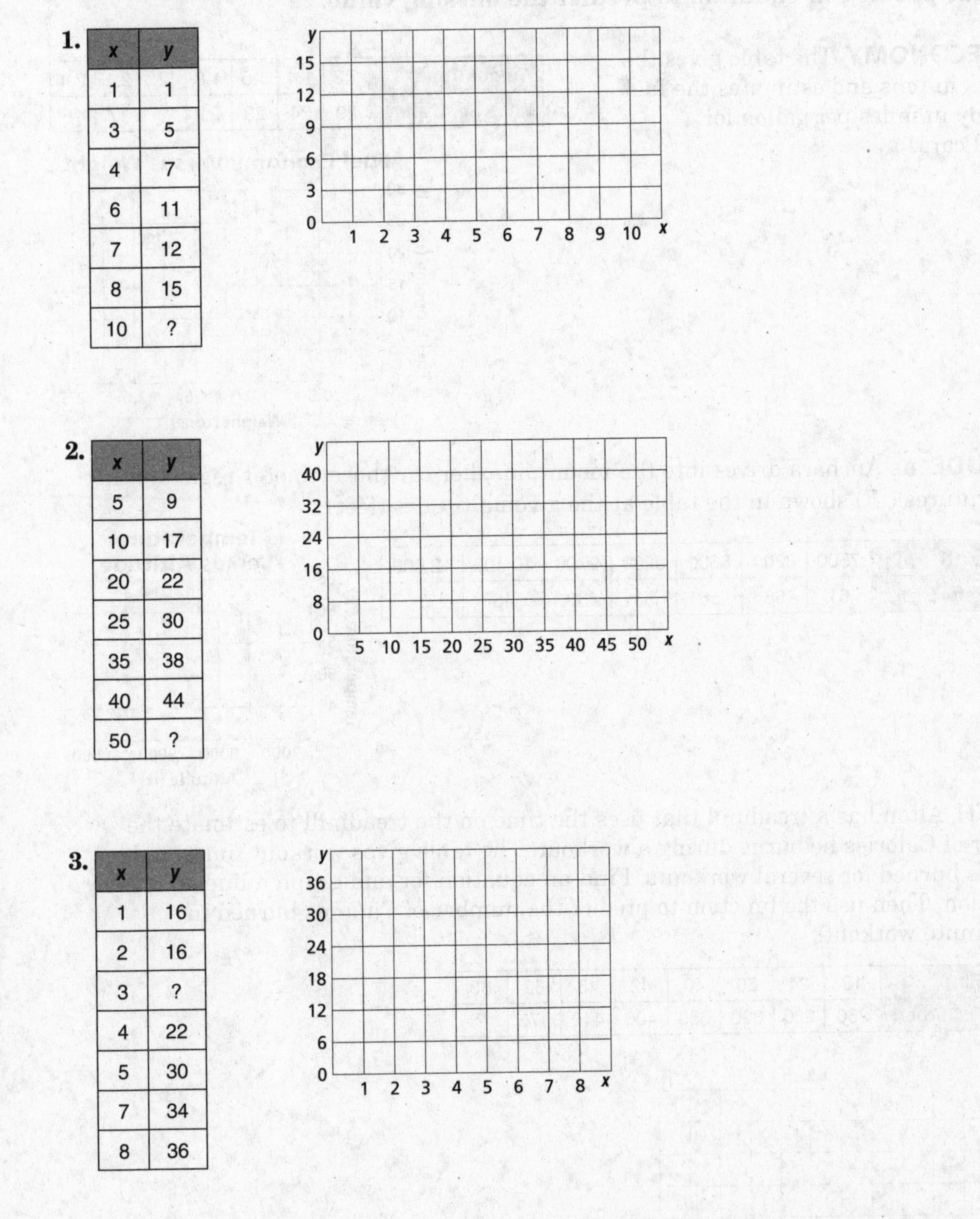

1.

x	y
1	1
3	5
4	7
6	11
7	12
8	15
10	?

2.

x	y
5	9
10	17
20	22
25	30
35	38
40	44
50	?

3.

x	y
1	16
2	16
3	?
4	22
5	30
7	34
8	36

NAME ______________________ DATE ____________ PERIOD ____________

2-5 Practice

Scatter Plots and Lines of Regression

For Exercises 1 and 2, complete parts a–c.

a. Make a scatter plot and a line of fit, and describe the correlation.

b. Use two ordered pairs to write a prediction equation.

c. Use your prediction equation to predict the missing value.

1. **FUEL ECONOMY** The table gives the weights in tons and estimates the fuel economy in miles per gallon for several cars.

Weight (tons)	1.3	1.4	1.5	1.8	2	2.1	2.4
Miles per Gallon	29	24	23	21	?	17	15

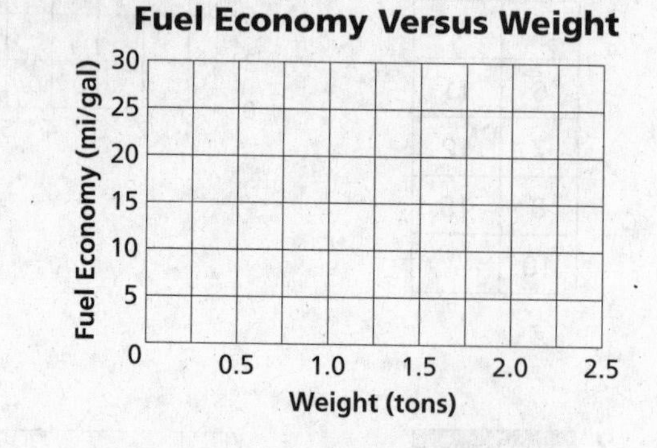

2. **ALTITUDE** As Anchara drives into the mountains, her car thermometer registers the temperatures (F) shown in the table at the given altitudes (feet).

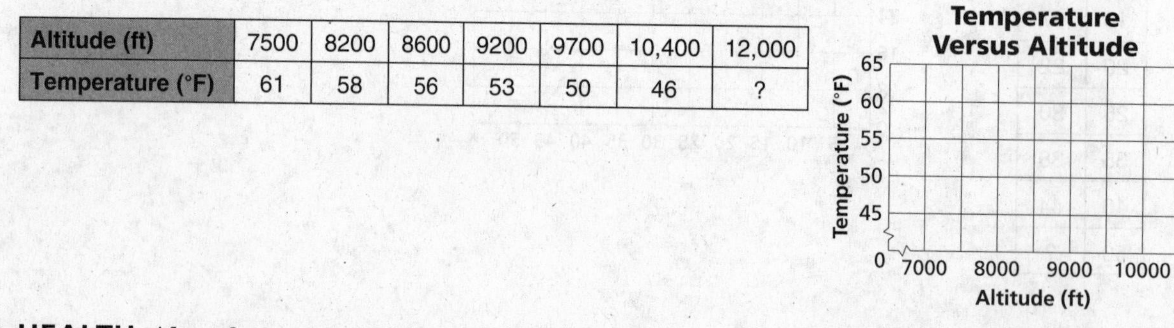

Altitude (ft)	7500	8200	8600	9200	9700	10,400	12,000
Temperature (°F)	61	58	56	53	50	46	?

3. **HEALTH** Alton has a treadmill that uses the time on the treadmill to estimate the number of Calories he burns during a workout. The table gives workout times and Calories burned for several workouts. Find an equation for and graph a line of regression. Then use the function to predict the number of Calories burned in a 60-minute workout.

Time (min)	18	24	30	40	42	48	52	60
Calories Burned	260	280	320	380	400	440	475	?

NAME ______________________ DATE ____________ PERIOD ________

2-6 Skills Practice

Special Functions

Graph each function. Identify the domain and range.

1. $f(x) = \begin{cases} -1 \text{ if } x \leq 0 \\ 2x \text{ if } 0 < x \leq 3 \\ 6 \text{ if } x > 3 \end{cases}$

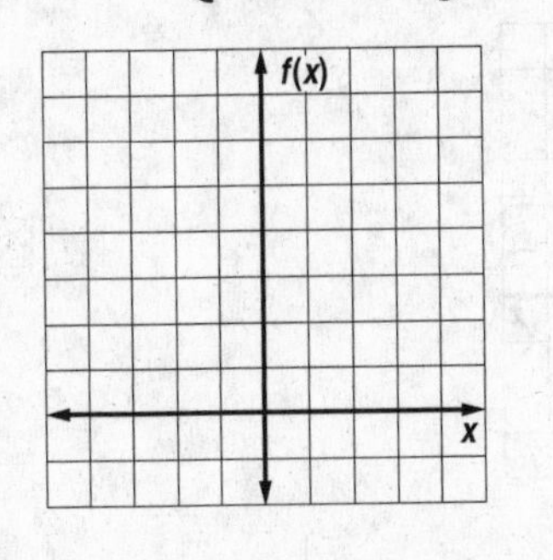

2. $f(x) = \begin{cases} -x \text{ if } x < 1 \\ 0 \text{ if } -1 \leq x \leq 1 \\ x \text{ if } x > 1 \end{cases}$

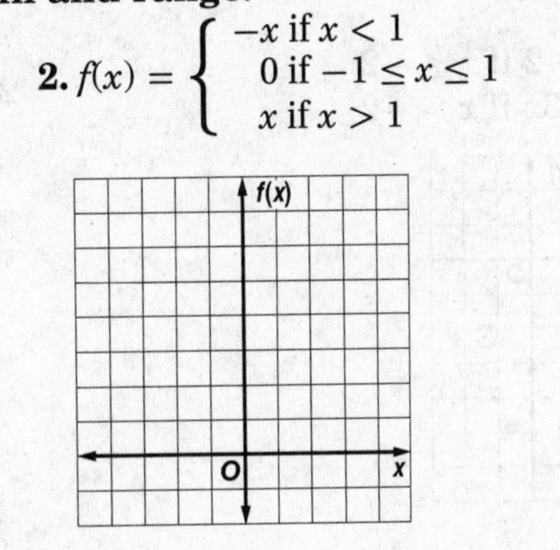

3. $f(x) = \begin{cases} x \text{ if } x < 0 \\ 2 \text{ if } x \geq 0 \end{cases}$

4. $h(x) = \begin{cases} 3 \text{ if } x < -1 \\ x + 1 \text{ if } x > 1 \end{cases}$

5. $f(x) = [\![x + 1]\!]$

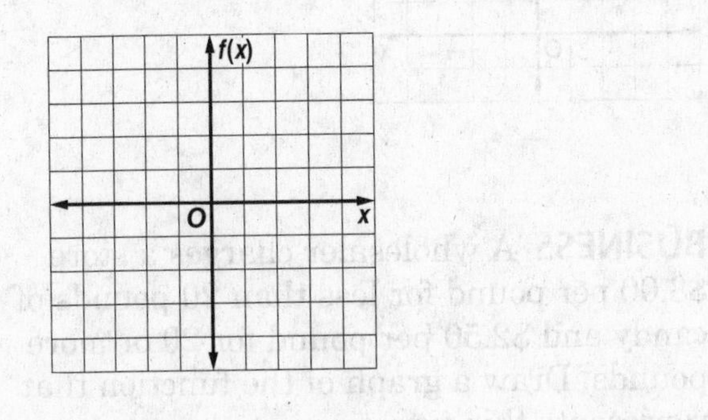

6. $f(x) = [\![x - 3]\!]$

7. $g(x) = 2|x|$

8. $f(x) = |x| + 1$

NAME ______________________________ DATE ____________ PERIOD ____________

2-6 Practice

Special Functions

Graph each function. Identify the domain and range.

1. $f(x) = \begin{cases} x + 2 \text{ if } x \le -2 \\ 3x \text{ if } x > -2 \end{cases}$

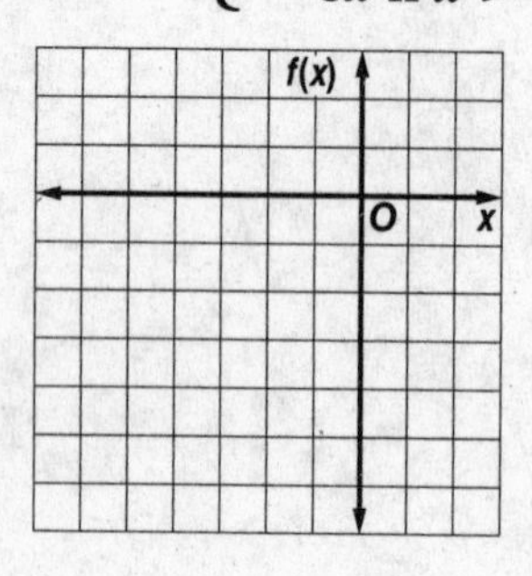

2. $h(x) = \begin{cases} 4 - x \text{ if } x > 0 \\ -2x - 2 \text{ if } x < 0 \end{cases}$

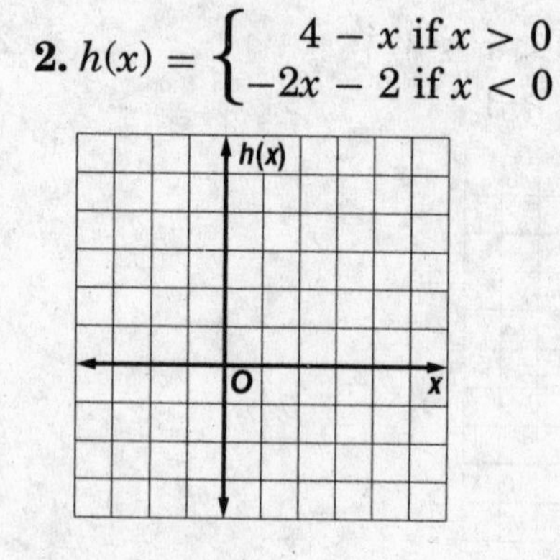

3. $f(x) = [\![0.5x]\!]$

f(x)
O
x

4. $f(x) = [\![x]\!] - 2$

f(x)
O
x

5. $g(x) = -2|x|$

g(x)
O
x

6. $f(x) = |x + 1|$

f(x)
O
x

7. BUSINESS *A Stitch in Time* charges $40 per hour or any fraction thereof for labor. Draw a graph of the step function that represents this situation.

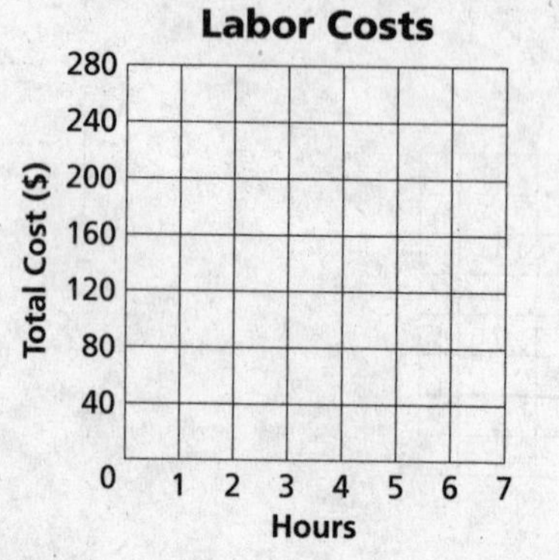

8. BUSINESS A wholesaler charges a store $3.00 per pound for less than 20 pounds of candy and $2.50 per pound for 20 or more pounds. Draw a graph of the function that represents this situation.

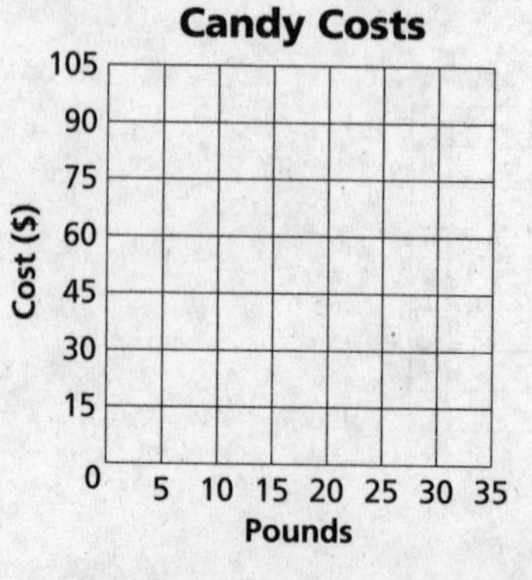

NAME ______________________ DATE __________ PERIOD __________

2-7 Skills Practice

Parent Functions and Transformation

Identify the type of function represented by each graph.

1.

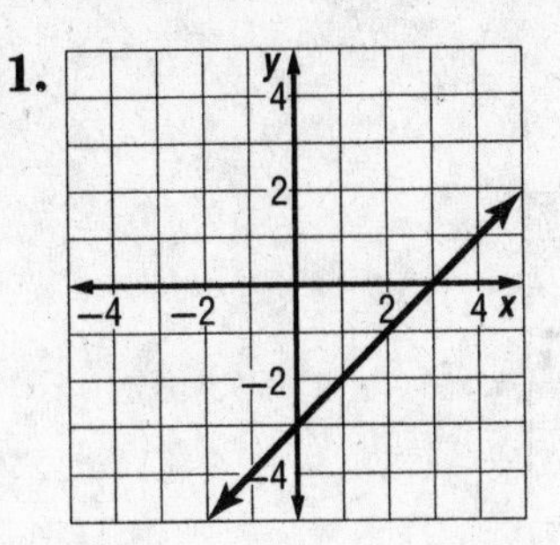

2.

Describe the translation in each equation. Then graph the function.

3. $y = |x| - 2$

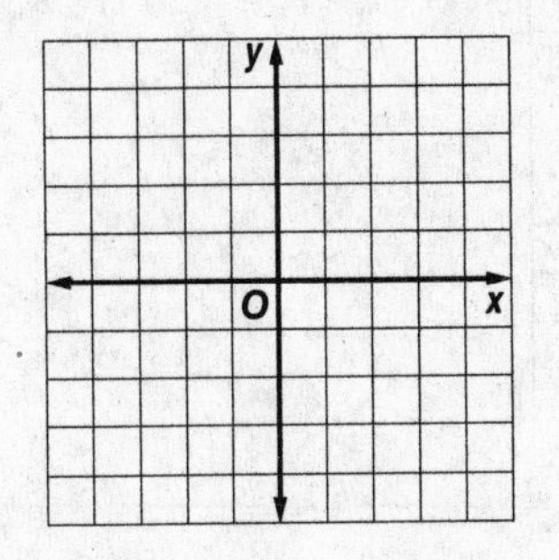

4. $y = (x + 1)^2$

Describe the reflection in each equation. Then graph the function.

5. $y = -x$

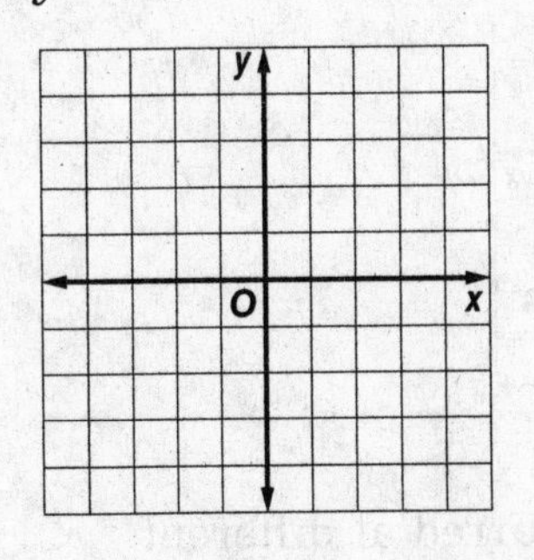

6. $y = -|x|$

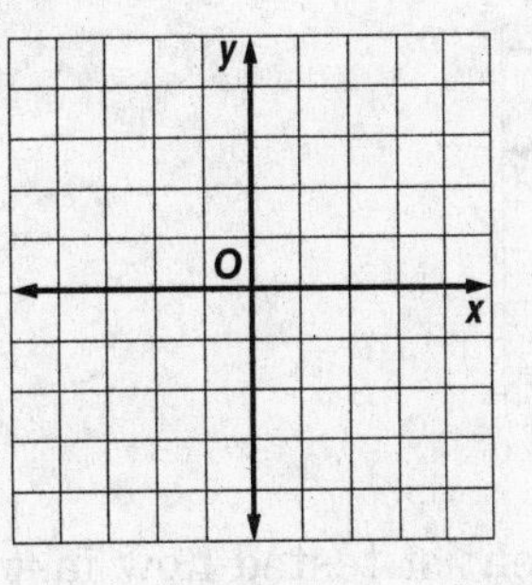

7. Biology A biologist plotted the data from his latest experiment and found that the graph of his data looked like this graph. What type of function relates the variables in the experiment?

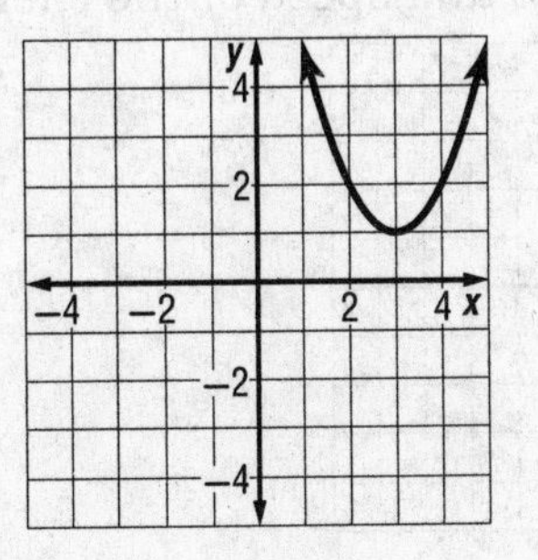

NAME ____________________ DATE ____________ PERIOD ____________

2-7 Practice

Parent Functions and Transformations

Describe the translation in each function. Then graph the function.

1. $y = x + 3$

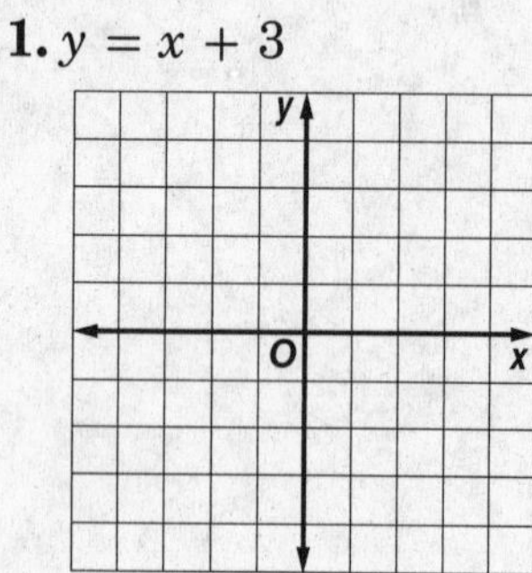

2. $y = x^2 - 3$

Describe the reflection in each function. Then graph the function.

3. $y = (-x)^2$

4. $y = -(3)$

Describe the dilation in each function. Then graph the function.

5. $y = |2x|$

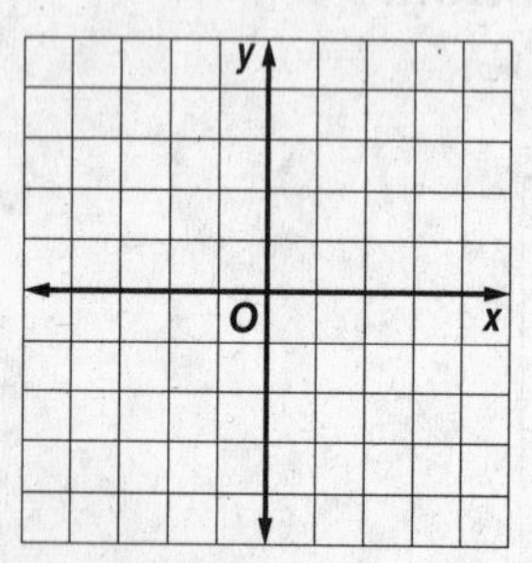

6. $4y = x^2$

7. CHEMISTRY A scientist tested how fast a chemical reaction occurred at different temperatures. The data made this graph. What type of function shows the relation of temperature and speed of the chemical reaction?

NAME ______________________ DATE __________ PERIOD __________

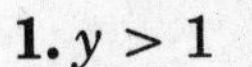

2-8 Skills Practice

Graphing Linear and Absolute Value Inequalities

Graph each inequality.

1. $y > 1$

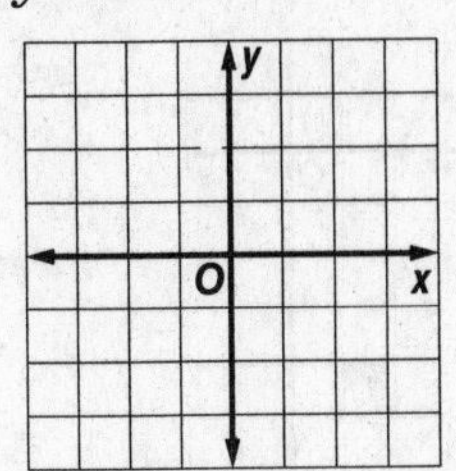

2. $y \leq x + 2$

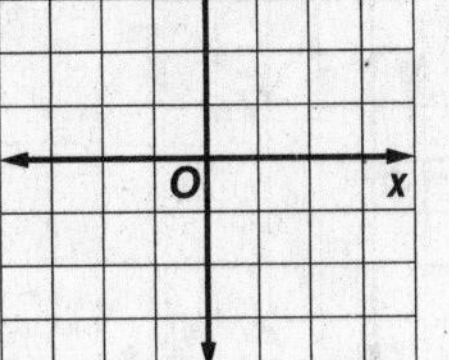

3. $x + y \leq 4$

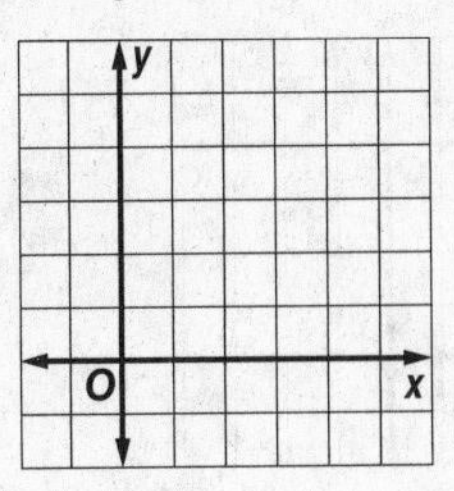

4. $x + 3 < y$

5. $2 - y < x$

6. $y \geq -x$

7. $x - y > -2$

8. $9x + 3y - 6 \leq 0$

9. $y + 1 \geq 2x$

10. $y - 7 \leq -9$

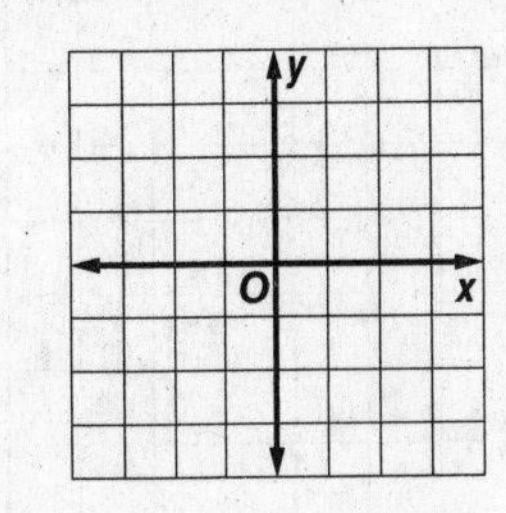

11. $x > -5$

12. $y > |x|$

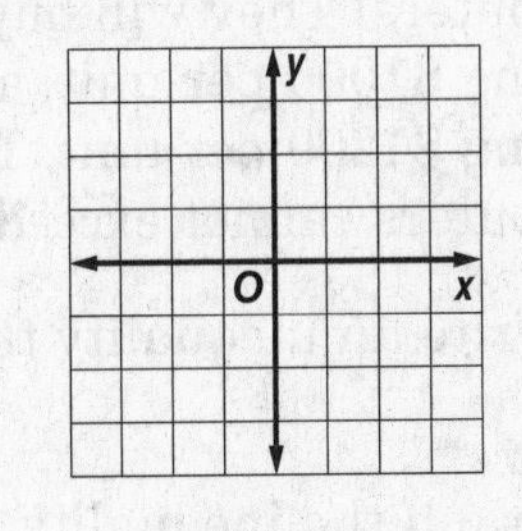

2-8 Practice

Graphing Linear and Absolute Value Inequalities

Graph each inequality.

1. $y \leq -3$

2. $x > 2$

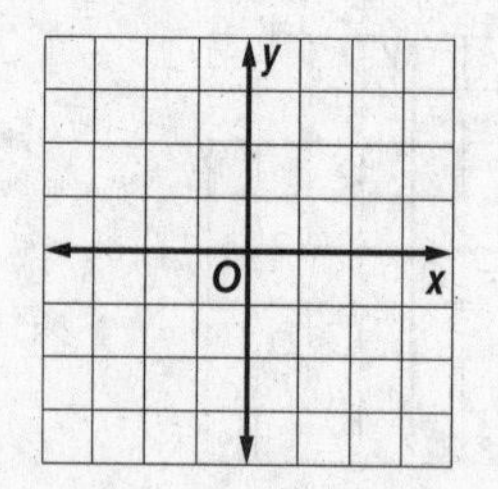

3. $x + y \leq -4$

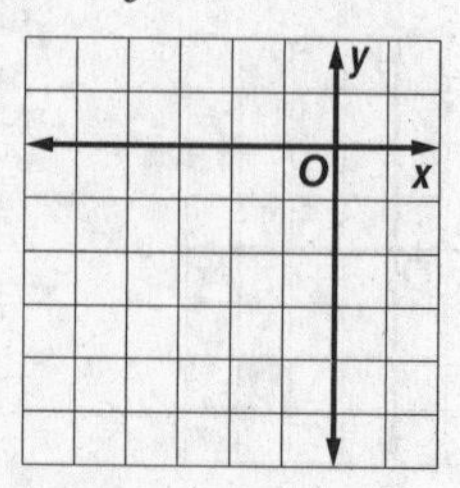

4. $y < -3x + 5$

5. $y < \frac{1}{2}x + 3$

6. $y - 1 \geq -x$

7. $x - 3y \leq 6$

8. $y > |x| - 1$

9. $y > -3|x + 1| - 2$

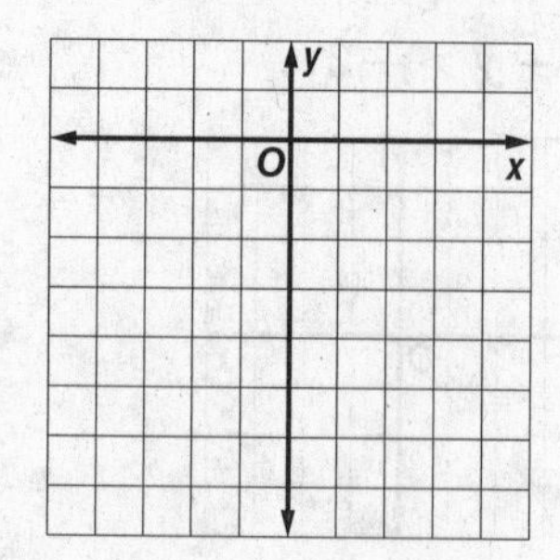

10. COMPUTERS A school system is buying new computers. They will buy desktop computers costing \$1000 per unit, and notebook computers costing \$1200 per unit. The total cost of the computers cannot exceed \$80,000.

a. Write an inequality that describes this situation.

b. Graph the inequality.

c. If the school wants to buy 50 of the desktop computers and 25 of the notebook computers, will they have enough money?

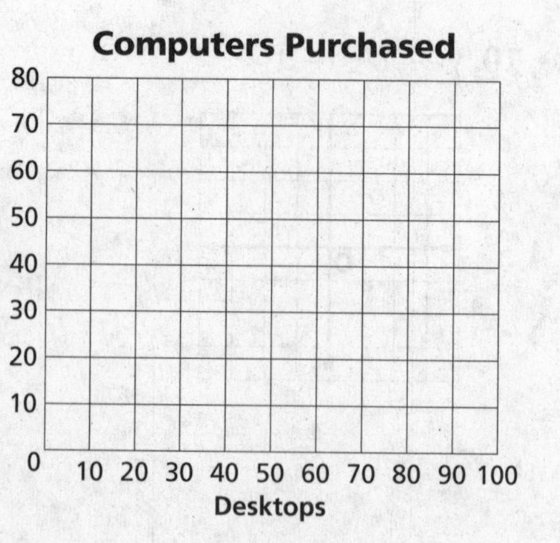

NAME ______________________ DATE ____________ PERIOD ____________

3-1 Skills Practice

Solving Systems of Equations By Graphing

Solve each system of equations by graphing.

1. $x = 2$
$y = 0$

2. $y = -3x + 6$
$y = 2x - 4$

3. $y = 4 - 3x$
$y = -\frac{1}{2}x - 1$

4. $y = 4 - x$
$y = x - 2$

5. $y = -2x + 2$
$y = \frac{1}{3}x - 5$

6. $y = x$
$y = -3x + 4$

7. $x + y = 3$
$x - y = 1$

8. $x - y = 4$
$2x - 5y = 8$

9. $3x - 2y = 4$
$2x - y = 1$

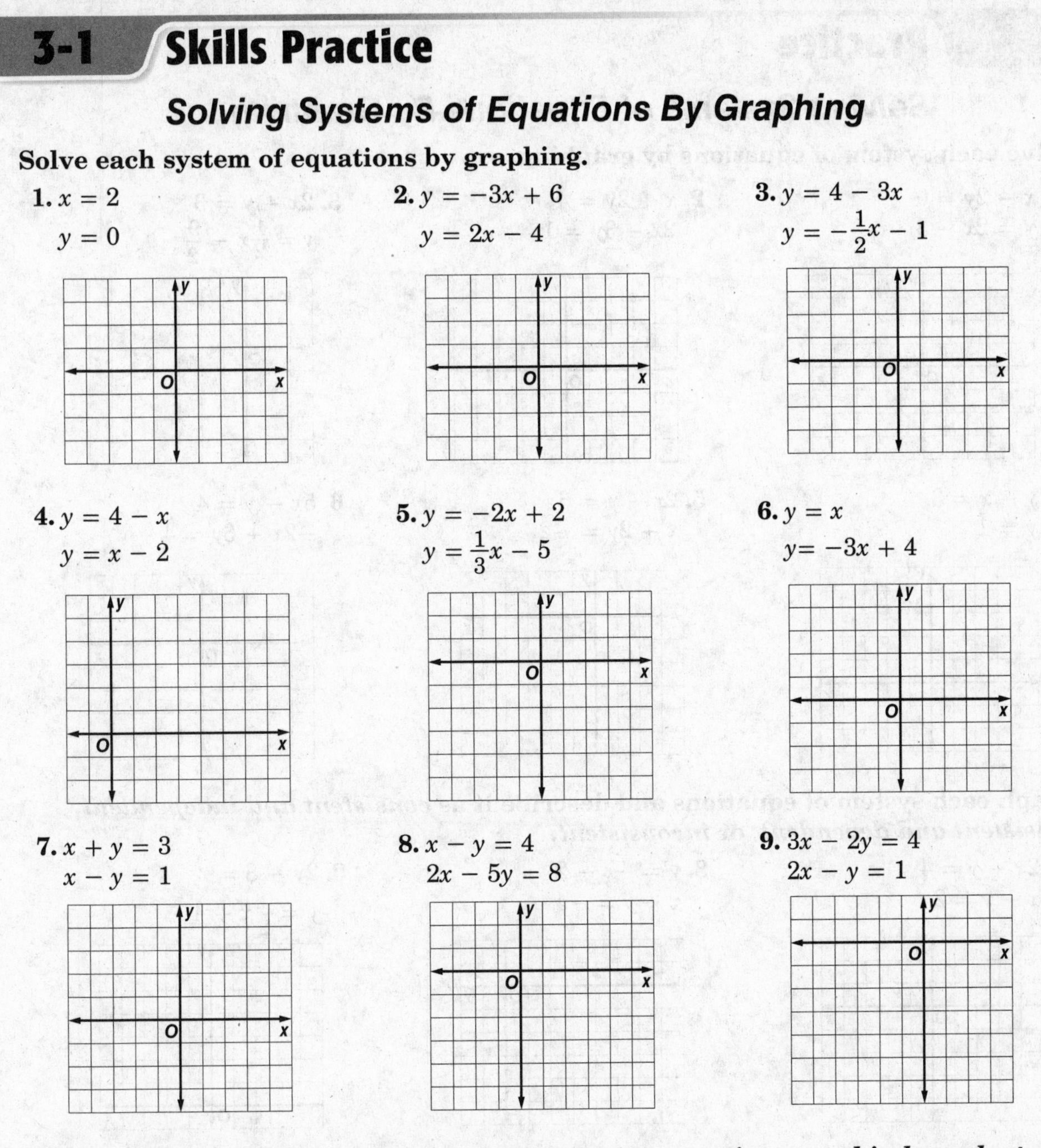

Graph each system of equations and describe it as *consistent and independent*, *consistent and dependent*, or *inconsistent*.

10. $y = -3x$
$y = -3x + 2$

11. $y = x - 5$
$-2x + 2y = -10$

12. $2x - 5y = 10$
$3x + y = 15$

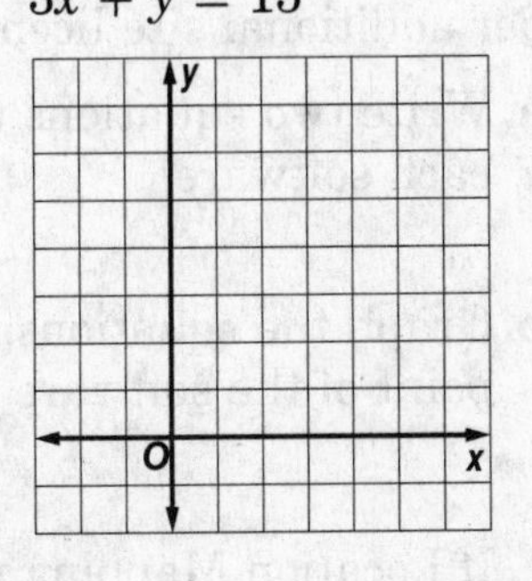

NAME ______________________ DATE ____________ PERIOD ________

3-1 Practice

Solving Systems of Equations By Graphing

Solve each system of equations by graphing.

1. $x - 2y = 0$
$y = 2x - 3$

2. $x + 2y = 4$
$2x - 3y = 1$

3. $2x + y = 3$
$y = \frac{1}{2}x - \frac{9}{2}$

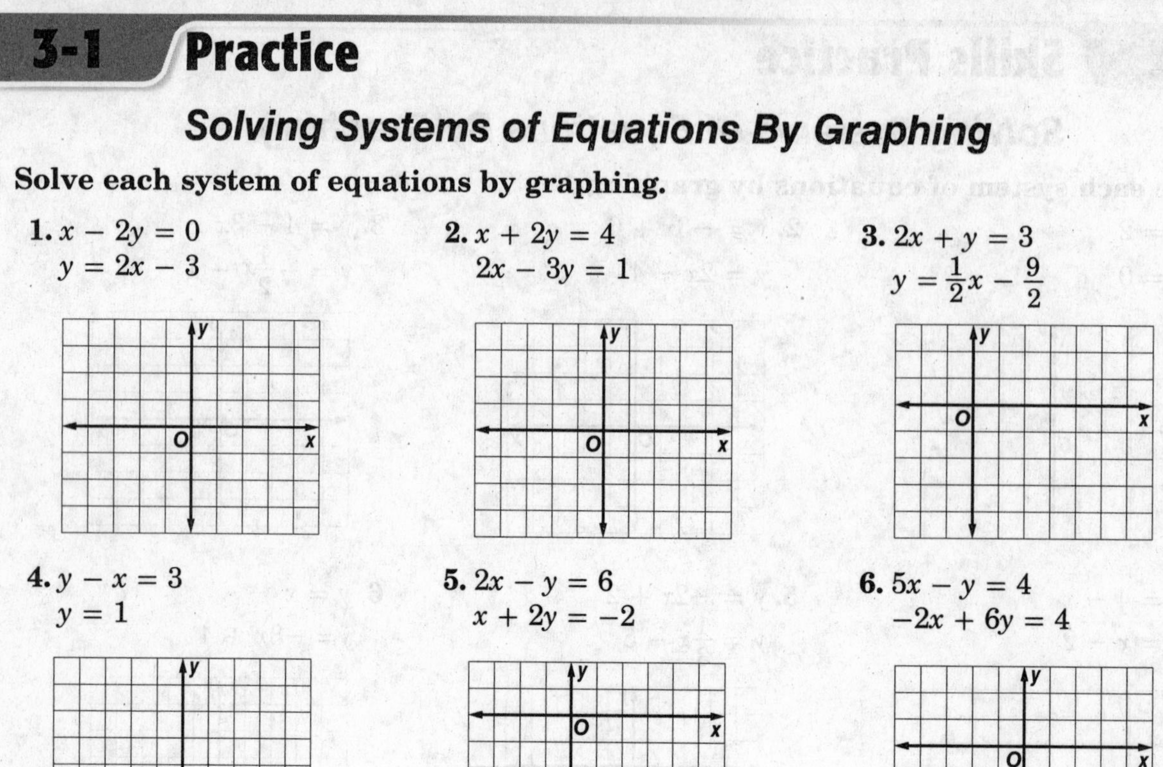

4. $y - x = 3$
$y = 1$

5. $2x - y = 6$
$x + 2y = -2$

6. $5x - y = 4$
$-2x + 6y = 4$

Graph each system of equations and describe it as *consistent and independent*, *consistent and dependent*, or *inconsistent*.

7. $2x - y = 4$
$x - y = 2$

8. $y = -x - 2$
$x + y = -4$

9. $2y - 8 = x$
$y = \frac{1}{2}x + 4$

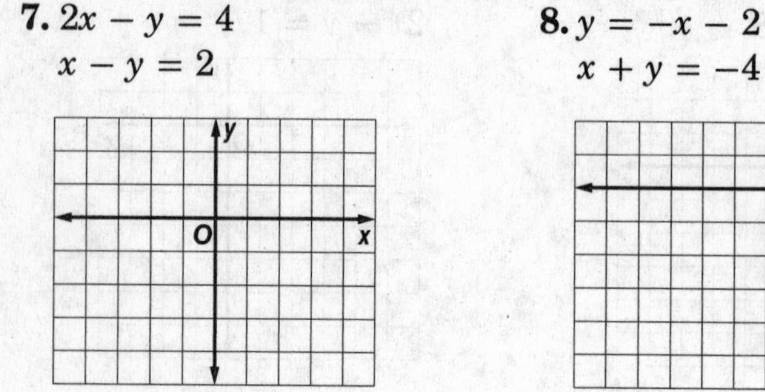

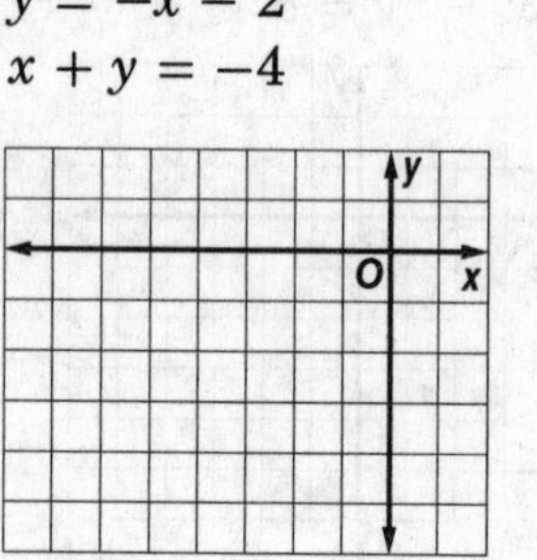

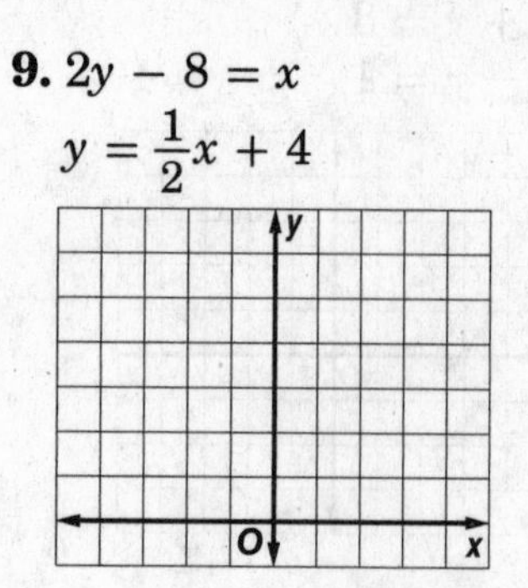

10. SOFTWARE Location Mapping needs new software. Software A costs $13,000 plus $500 per additional site license. Software B costs $2500 plus $1200 per additional site license.

a. Write two equations that represent the cost of each software.

b. Graph the equations. Estimate the break-even point of the software costs.

c. If Location Mapping plans to buy 10 additional site licenses, which software will cost less?

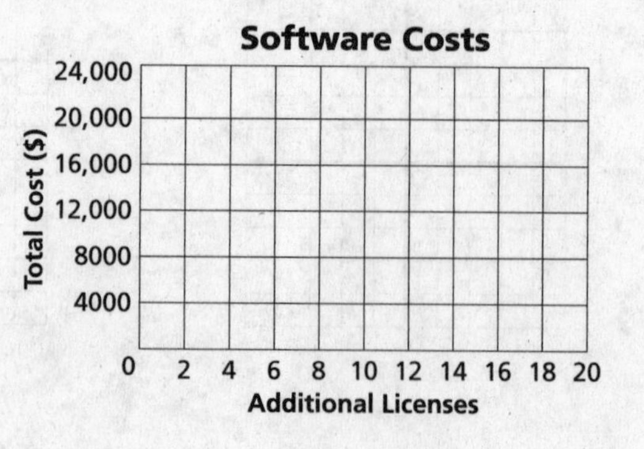

3-2 Skills Practice

Solving Systems of Equations Algebraically

Solve each system of equations by using substitution.

1. $a + b = 20$
$a - b = -4$

2. $x + 3y = -3$
$4x + 3y = 6$

3. $w - z = 1$
$2w + 3z = 12$

4. $3r + t = 5$
$2r - t = 5$

5. $2b + 3c = -4$
$b + c = 3$

6. $x - y = -5$
$3x + 4y = 13$

Solve each system of equations by using elimination.

7. $2t - u = 17$
$3t + u = 8$

8. $2j - k = 3$
$3j + k = 2$

9. $3c - 2d = 2$
$3c + 4d = 50$

10. $2f + 3g = 9$
$f - g = 2$

11. $-2x + y = -1$
$x + 2y = 3$

12. $2x - y = 12$
$2x - y = 6$

Solve each system of equations.

13. $-r + t = 5$
$-2r + t = 4$

14. $2x - y = -5$
$4x + y = 2$

15. $x - 3y = -12$
$2x + y = 11$

16. $2p - 3r = 6$
$-2p + 3r = -6$

17. $6w - 8z = 16$
$3w - 4z = 8$

18. $c + d = 6$
$c - d = 0$

19. $2u + 4x = -6$
$u + 2x = 3$

20. $3a + b = -1$
$-3a + b = 5$

21. $2x + y = 6$
$3x - 2y = 16$

22. $3y - z = -6$
$-3y - z = 6$

23. $c + 2d = -2$
$-2c - 5d = 3$

24. $3r - 2t = 1$
$2r - 3t = 9$

25. The sum of two numbers is 12. The difference of the same two numbers is -4. Find the numbers.

26. Twice a number minus a second number is -1. Twice the second number added to three times the first number is 9. Find the two numbers.

3-2 Practice

Solving Systems of Equations Algebraically

Solve each system of equations by using substitution.

1. $2x + y = 4$
$3x + 2y = 1$

2. $x - 3y = 9$
$x + 2y = -1$

3. $g + 3h = 8$
$\frac{1}{3}g + h = 9$

4. $2a - 4b = 6$
$-a + 2b = -3$

5. $2m + n = 6$
$5m + 6n = 1$

6. $4x - 3y = -6$
$-x - 2y = 7$

7. $u - 2v = \frac{1}{2}$
$-u + 2v = 5$

8. $x - 3y = 16$
$4x - y = 9$

9. $w + 3z = 1$
$3w - 5z = -4$

Solve each system of equations by using elimination.

10. $2r + t = 5$
$3r - t = 20$

11. $2m - n = -1$
$3m + 2n = 30$

12. $6x + 3y = 6$
$8x + 5y = 12$

13. $3j - k = 10$
$4j - k = 16$

14. $2x - y = -4$
$-4x + 2y = 6$

15. $2g + h = 6$
$3g - 2h = 16$

16. $2t + 4v = 6$
$-t - 2v = -3$

17. $3x - 2y = 12$
$2x + \frac{2}{3}y = 14$

18. $\frac{1}{2}x + 3y = 11$
$8x - 5y = 17$

Solve each system of equations.

19. $8x + 3y = -5$
$10x + 6y = -13$

20. $8q - 15r = -40$
$4q + 2r = 56$

21. $3x - 4y = 12$
$\frac{1}{3}x - \frac{4}{9}y = \frac{4}{3}$

22. $4b - 2d = 5$
$-2b + d = 1$

23. $x + 3y = 4$
$x = 1$

24. $4m - 2p = 0$
$-3m + 9p = 5$

25. $5g + 4k = 10$
$-3g - 5k = 7$

26. $0.5x + 2y = 5$
$x - 2y = -8$

27. $h - z = 3$
$-3h + 3z = 6$

28. SPORTS Last year the volleyball team paid \$5 per pair for socks and \$17 per pair for shorts on a total purchase of \$315. This year they spent \$342 to buy the same number of pairs of socks and shorts because the socks now cost \$6 a pair and the shorts cost \$18.

a. Write a system of two equations that represents the number of pairs of socks and shorts bought each year.

b. How many pairs of socks and shorts did the team buy each year?

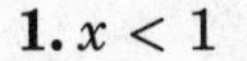

3-3 Skills Practice

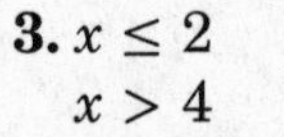

Solving Systems of Inequalities by Graphing

Solve each system of inequalities by graphing.

1. $x < 1$
$y \geq -1$

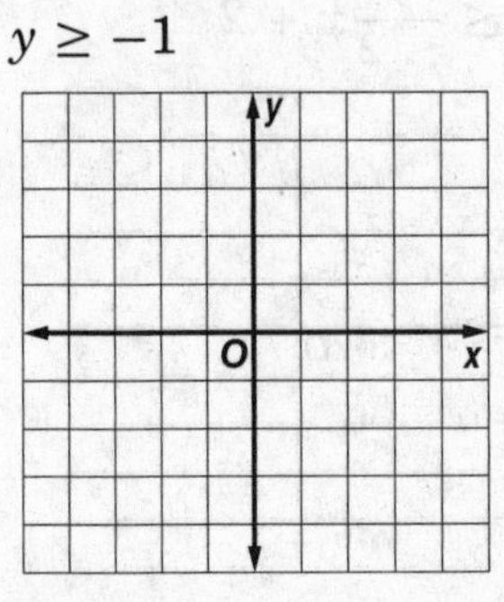

2. $x \geq -3$
$y \geq -3$

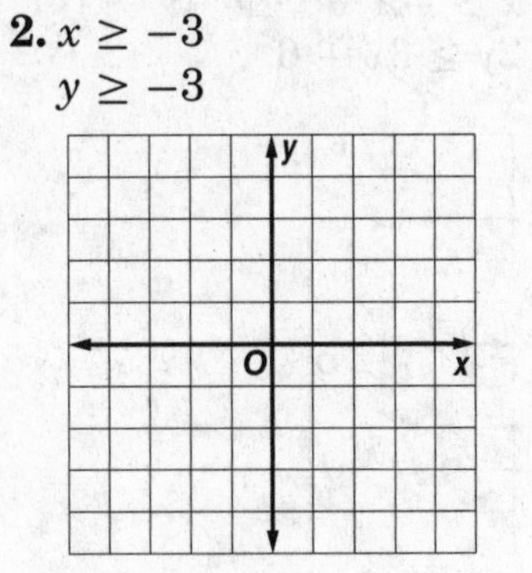

3. $x \leq 2$
$x > 4$

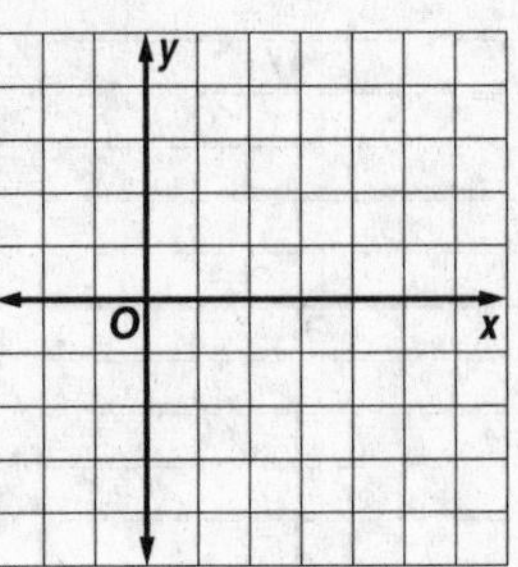

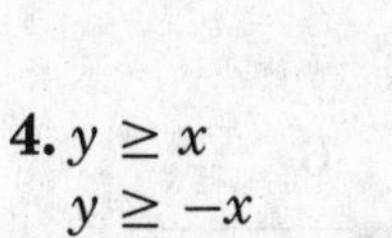

4. $y \geq x$
$y \geq -x$

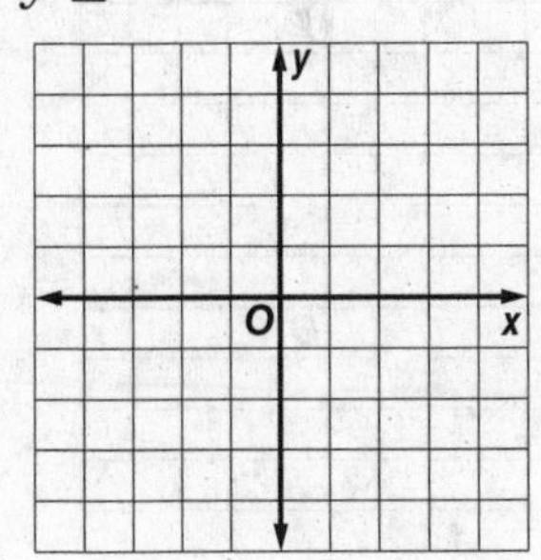

5. $y < -4x$
$y \geq 3x - 2$

6. $x - y \geq -1$
$3x - y \leq 4$

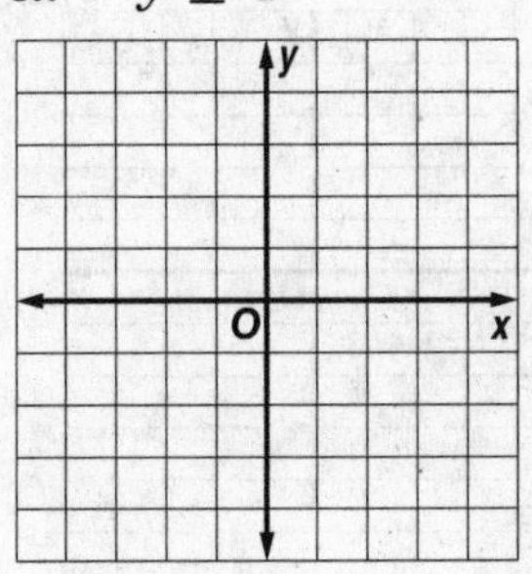

7. $y < 3$
$x + 2y < 12$

8. $y < -2x + 3$
$y \leq x - 2$

9. $x - y \leq 4$
$2x + y < 4$

Find the coordinates of the vertices of the triangle formed by each system of inequalities.

10. $y \leq 0$
$x \leq 0$
$y \geq -x - 1$

11. $y \leq 3 - x$
$y \geq 3$
$x \geq -5$

12. $x \geq -2$
$y \geq x - 2$
$x + y \leq 2$

NAME ______________________ DATE ____________ PERIOD ________

3-3 Practice

Solving Systems of Inequalities by Graphing

Solve each system of inequalities by graphing.

1. $y + 1 < -x$
$y \geq 1$

2. $x > -2$
$2y \geq 3x + 6$

3. $y \leq 2x - 3$
$y \leq -\frac{1}{2}x + 2$

4. $x + y > -2$
$3x - y \geq -2$

5. $|y| \leq 1$
$y < x - 1$

6. $3y > 4x$
$2x - 3y > -6$

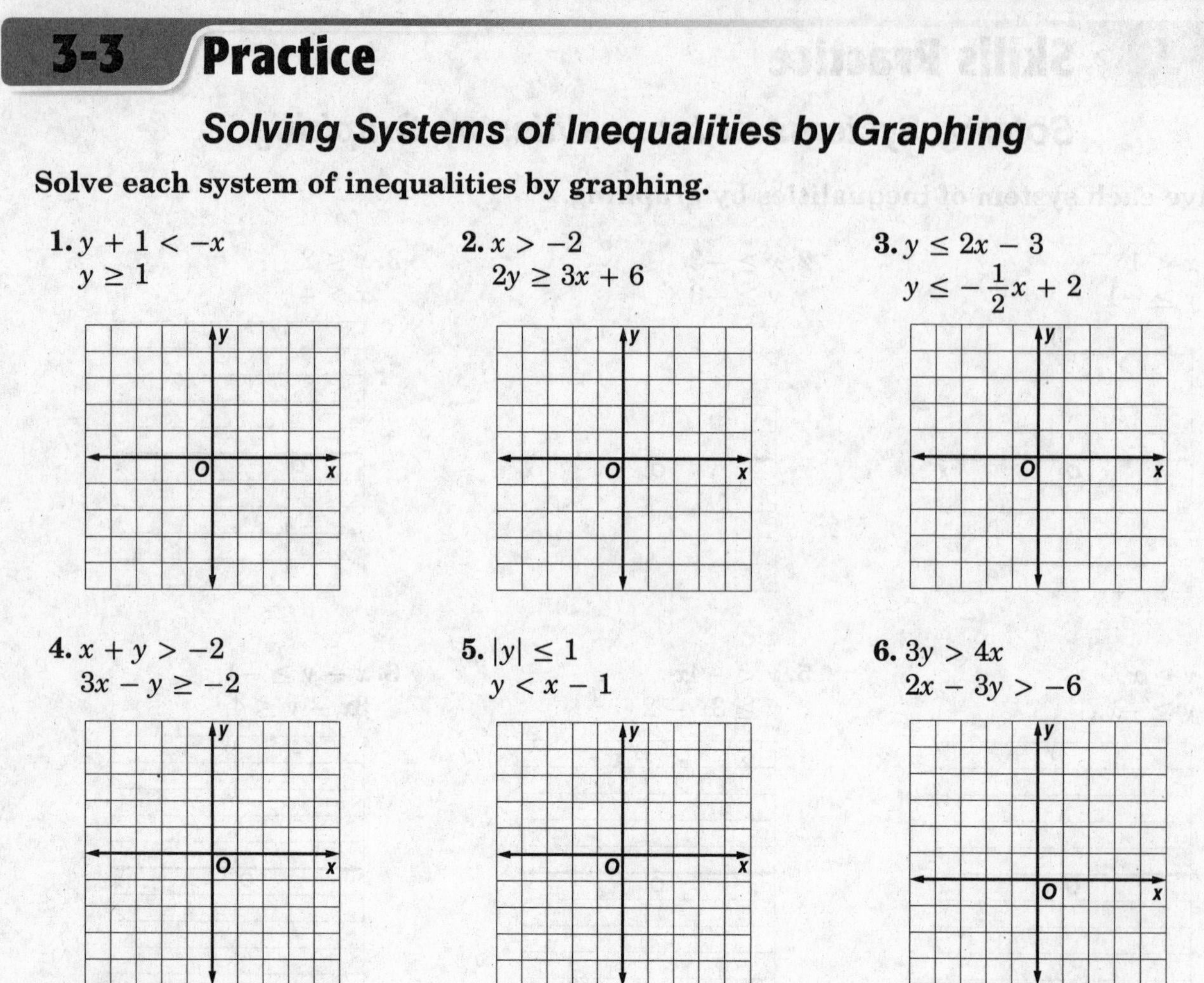

Find the coordinates of the vertices of the triangle formed by each system of inequalities.

7. $y \geq 1 - x$
$y \leq x - 1$
$x \leq 3$

8. $x - y \leq 2$
$x + y \leq 2$
$x \geq -2$

9. $y \geq 2x - 2$
$2x + 3y \geq 6$
$y < 4$

10. DRAMA The drama club is selling tickets to its play. An adult ticket costs $15 and a student ticket costs $11. The auditorium will seat 300 ticket-holders. The drama club wants to collect at least $3630 from ticket sales.

a. Write and graph a system of four inequalities that describe how many of each type of ticket the club must sell to meets its goal.

b. List three different combinations of tickets sold that satisfy the inequalities.

NAME _______________ DATE _______________ PERIOD _______________

3-4 Skills Practice

Optimization with Linear Programming

Graph each system of inequalities. Name the coordinates of the vertices of the feasible region. Find the maximum and minimum values of the given function for this region.

1. $x \geq 2$
$x \leq 5$
$y \geq 1$
$y \leq 4$
$f(x, y) = x + y$

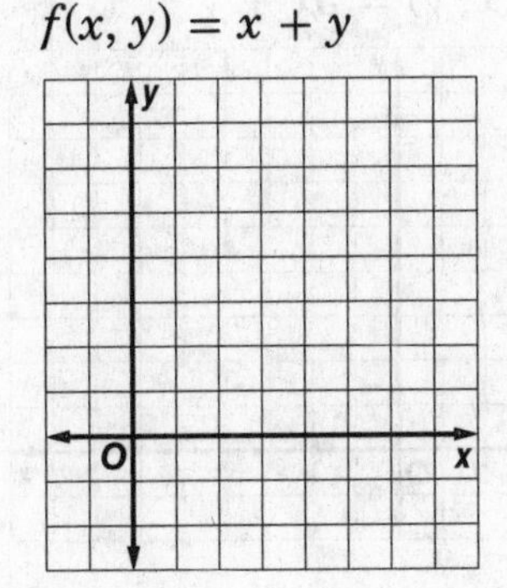

2. $x \geq 1$
$y \leq 6$
$y \geq x - 2$
$f(x, y) = x - y$

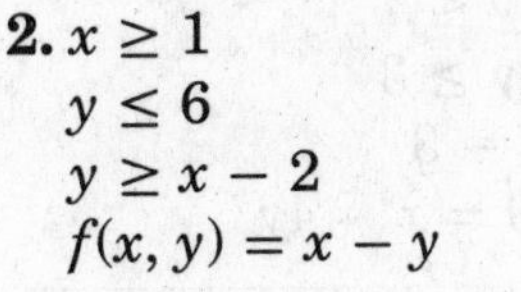

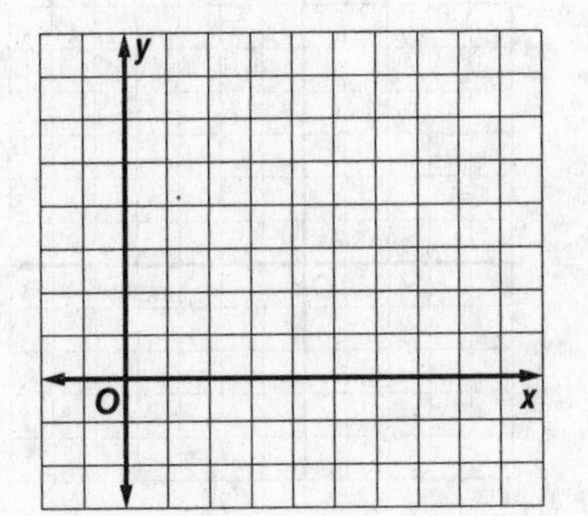

3. $x \geq 0$
$y \geq 0$
$y \leq 7 - x$
$f(x, y) = 3x + y$

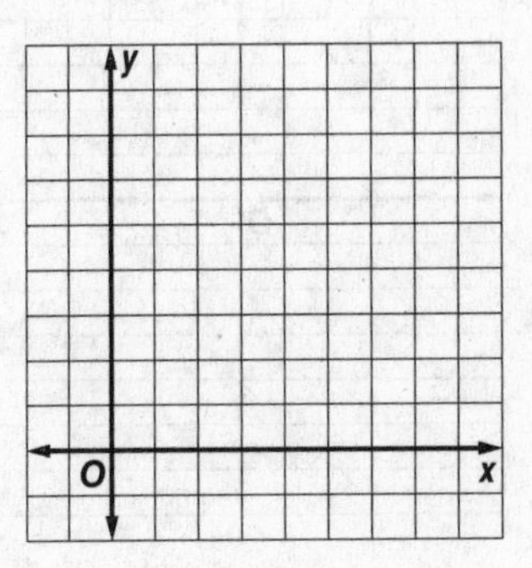

4. $x \geq -1$
$x + y \leq 6$
$f(x, y) = x + 2y$

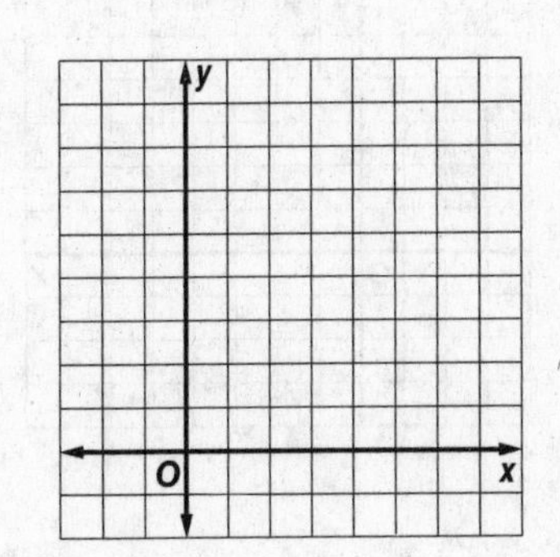

5. $y \leq 2x$
$y \geq 6 - x$
$y \leq 6$
$f(x, y) = 4x + 3y$

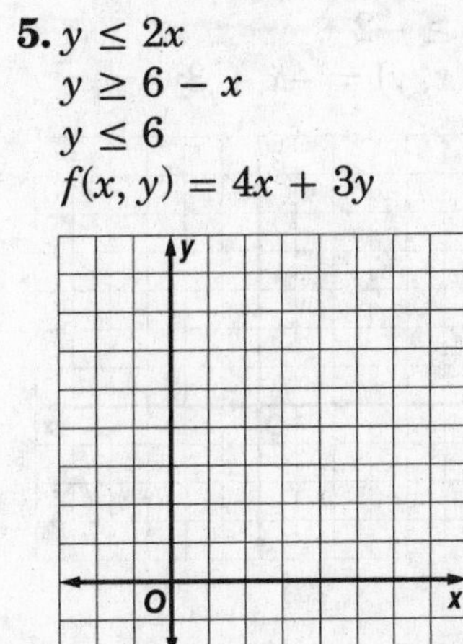

6. $y \geq -x - 2$
$y \geq 3x + 2$
$y \leq x + 4$
$f(x, y) = -3x + 5y$

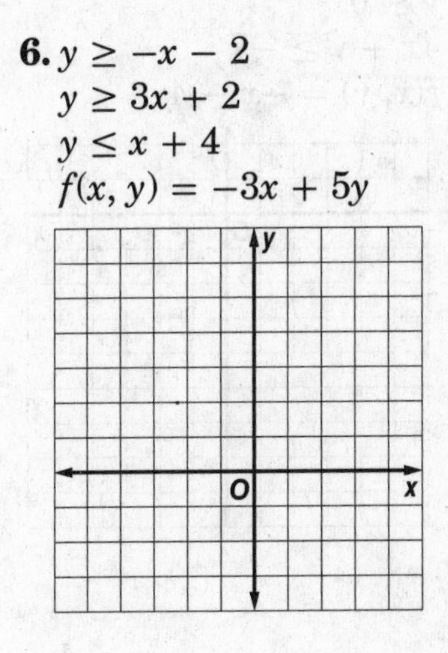

7. MANUFACTURING A backpack manufacturer produces an internal frame pack and an external frame pack. Let x represent the number of internal frame packs produced in one hour and let y represent the number of external frame packs produced in one hour. Then the inequalities $x + 3y \leq 18$, $2x + y \leq 16$, $x \geq 0$, and $y \geq 0$ describe the constraints for manufacturing both packs. Use the profit function $f(x, y) = 50x + 80y$ and the constraints given to determine the maximum profit for manufacturing both backpacks for the given constraints.

3-4 Practice

Optimization with Linear Programming

Graph each system of inequalities. Name the coordinates of the vertices of the feasible region. Find the maximum and minimum values of the given function for this region.

1. $2x - 4 \le y$
$-2x - 4 \le y$
$y \le 2$
$f(x, y) = -2x + y$

2. $3x - y \le 7$
$2x - y \ge 3$
$y \ge x - 3$
$f(x, y) = x - 4y$

3. $x \ge 0$
$y \ge 0$
$y \le 6$
$y \le -3x + 15$
$f(x, y) = 3x + y$

4. $x \le 0$
$y \le 0$
$4x + y \ge -7$
$f(x, y) = -x - 4y$

5. $y \le 3x + 6$
$4y + 3x \le 3$
$x \ge -2$
$f(x, y) = -x + 3y$

6. $2x + 3y \ge 6$
$2x - y \le 2$
$x \ge 0$
$y \ge 0$
$f(x, y) = x + 4y + 3$

7. PRODUCTION A glass blower can form 8 simple vases or 2 elaborate vases in an hour. In a work shift of no more than 8 hours, the worker must form at least 40 vases.

a. Let s represent the hours forming simple vases and e the hours forming elaborate vases. Write a system of inequalities involving the time spent on each type of vase.

b. If the glass blower makes a profit of \$30 per hour worked on the simple vases and \$35 per hour worked on the elaborate vases, write a function for the total profit on the vases.

c. Find the number of hours the worker should spend on each type of vase to maximize profit. What is that profit?

NAME ______________________ DATE __________ PERIOD ________

3-5 Skills Practice

Systems of Equations in Three Variables

Solve each system of equations.

1. $2a + c = -10$
$b - c = 15$
$a - 2b + c = -5$

2. $x + y + z = 3$
$13x + 2z = 2$
$-x - 5z = -5$

3. $2x + 5y + 2z = 6$
$5x - 7y = -29$
$z = 1$

4. $x + 4y - z = 1$
$3x - y + 8z = 0$
$x + 4y - z = 10$

5. $-2z = -6$
$2x + 3y - z = -2$
$x + 2y + 3z = 9$

6. $3x - 2y + 2z = -2$
$x + 6y - 2z = -2$
$x + 2y = 0$

7. $-x - 5z = -5$
$y - 3x = 0$
$13x + 2z = 2$

8. $-3x + 2z = 1$
$4x + y - 2z = -6$
$x + y + 4z = 3$

9. $x - y + 3z = 3$
$-2x + 2y - 6z = 6$
$y - 5z = -3$

10. $5x + 3y + z = 4$
$3x + 2y = 0$
$2x - y + 3z = 8$

11. $2x + 2y + 2z = -2$
$2x + 3y + 2z = 4$
$x + y + z = -1$

12. $x + 2y - z = 4$
$3x - y + 2z = 3$
$-x + 3y + z = 6$

13. $3x - 2y + z = 1$
$-x + y - z = 2$
$5x + 2y + 10z = 39$

14. $3x - 5y + 2z = -12$
$x + 4y - 2z = 8$
$-3x + 5y - 2z = 12$

15. $2x + y + 3z = -2$
$x - y - z = -3$
$3x - 2y + 3z = -12$

16. $2x - 4y + 3z = 0$
$x - 2y - 5z = 13$
$5x + 3y - 2z = 19$

17. $-2x + y + 2z = 2$
$3x + 3y + z = 0$
$x + y + z = 2$

18. $x - 2y + 2z = -1$
$x + 2y - z = 6$
$-3x + 6y - 6z = 3$

19. The sum of three numbers is 18. The sum of the first and second numbers is 15, and the first number is 3 times the third number. Find the numbers.

NAME ______________________ DATE ____________ PERIOD ____________

3-5 Practice

Systems of Equations in Three Variables

Solve each system of equations.

1. $2x - y + 2z = 15$
$-x + y + z = 3$
$3x - y + 2z = 18$

2. $x - 4y + 3z = -27$
$2x + 2y - 3z = 22$
$4z = -16$

3. $a + b = 3$
$-b + c = 3$
$a + 2c = 10$

4. $3m - 2n + 4p = 15$
$m - n + p = 3$
$m + 4n - 5p = 0$

5. $2g + 3h - 8j = 10$
$g - 4h = 1$
$-2g - 3h + 8j = 5$

6. $2x + y - z = -8$
$4x - y + 2z = -3$
$-3x + y + 2z = 5$

7. $2x - 5y + z = 5$
$3x + 2y - z = 17$
$4x - 3y + 2z = 17$

8. $2x + 3y + 4z = 2$
$5x - 2y + 3z = 0$
$x - 5y - 2z = -4$

9. $p + 4r = -7$
$p - 3q = -8$
$q + r = 1$

10. $4x + 4y - 2z = 8$
$3x - 5y + 3z = 0$
$2x + 2y - z = 4$

11. $d + 3g + h = 0$
$-d + 2g + h = -1$
$4d + g - h = 1$

12. $4x + y + 5z = -9$
$x - 4y - 2z = -2$
$2x + 3y - 2z = 21$

13. $5x + 9y + z = 20$
$2x - y - z = -21$
$5x + 2y + 2z = -21$

14. $2x + y - 3z = -3$
$3x + 2y + 4z = 5$
$-6x - 3y + 9z = 9$

15. $3x + 3y + z = 10$
$5x + 2y + 2z = 7$
$3x - 2y + 3z = -9$

16. $2u + v + w = 2$
$-3u + 2v + 3w = 7$
$-u - v + 2w = 7$

17. $x + 5y - 3z = -18$
$3x - 2y + 5z = 22$
$-2x - 3y + 8z = 28$

18. $x - 2y + z = -1$
$-x + 2y - z = 6$
$-4y + 2z = 1$

19. $2x - 2y - 4z = -2$
$3x - 3y - 6z = -3$
$-2x + 3y + z = 7$

20. $x - y + 9z = -27$
$2x - 4y - z = -1$
$3x + 6y - 3z = 27$

21. $2x - 5y - 3z = 7$
$-4x + 10y + 2z = 6$
$6x - 15y - z = -19$

22. The sum of three numbers is 6. The third number is the sum of the first and second numbers. The first number is one more than the third number. Find the numbers.

23. The sum of three numbers is −4. The second number decreased by the third is equal to the first. The sum of the first and second numbers is −5. Find the numbers.

24. SPORTS Alexandria High School scored 37 points in a football game. Six points are awarded for each touchdown. After each touchdown, the team can earn one point for the extra kick or two points for a 2-point conversion. The team scored one fewer 2-point conversions than extra kicks. The team scored 10 times during the game. How many touchdowns were made during the game?

NAME ______________________ DATE ____________ PERIOD ____________

4-1 Skills Practice

Introduction to Matrices

State the dimensions of each matrix.

1. $\begin{bmatrix} 3 & 2 & 4 \\ -1 & 4 & 0 \end{bmatrix}$

2. $[0 \ \ 15]$

3. $\begin{bmatrix} 3 & 2 \\ 1 & 8 \end{bmatrix}$

4. $\begin{bmatrix} 6 & 1 & 2 \\ -3 & 4 & 5 \\ -2 & 7 & 9 \end{bmatrix}$

5. $\begin{bmatrix} 9 & 3 & -3 & -6 \\ 3 & 4 & -4 & 5 \end{bmatrix}$

6. $\begin{bmatrix} -1 \\ -1 \\ -1 \\ -3 \end{bmatrix}$

Identify each element for the following matrices.

$A = \begin{bmatrix} 9 & 6 & 7 \\ 2 & 5 & 0 \\ 10 & 3 & 11 \end{bmatrix}$ $B = \begin{bmatrix} 5 & -2 & 4 & 3 \\ 0 & 8 & 12 & -1 \end{bmatrix}$ $C = \begin{bmatrix} 8 & 1 & 6 \\ 7 & 0 & 2 \\ 4 & 9 & 5 \\ 3 & 12 & 10 \end{bmatrix}$

7. b_{22}

8. c_{42}

9. b_{11}

10. a_{33}

11. c_{41}

12. a_{21}

13. c_{33}

14. b_{13}

15. a_{12}

4-1 Practice

Introduction to Matrices

State the dimensions of each matrix.

1. $[-3 \quad -3 \quad 7]$

2. $\begin{bmatrix} 5 & 8 & -1 \\ -2 & 1 & 8 \end{bmatrix}$

3. $\begin{bmatrix} -2 & 2 & -2 & 3 \\ 5 & 16 & 0 & 0 \\ 4 & 7 & -1 & 4 \end{bmatrix}$

Identify each element for the following matrices.

$A = \begin{bmatrix} 4 & 7 & 0 \\ 9 & 8 & -4 \\ 3 & 0 & 5 \\ -1 & 2 & 6 \end{bmatrix}$, $B = \begin{bmatrix} 2 & 6 & -1 & 0 \\ 9 & 5 & 7 & 2 \end{bmatrix}$.

4. b_{23}

5. a_{42}

6. b_{11}

7. a_{32}

8. b_{14}

9. a_{23}

10. TICKET PRICES The table at the right gives ticket prices for a concert. Write a 2 × 3 matrix that represents the cost of a ticket.

	Child	Student	Adult
Cost Purchased in Advance	$6	$12	$18
Cost Purchased at the Door	$8	$15	$22

11. CONSTRUCTION During each of the last three weeks, a road-building crew has used three truckloads of gravel. The table at the right shows the amount of gravel in each load.

Week 1		Week 2		Week 3	
Load 1	40 tons	Load 1	40 tons	Load 1	32 tons
Load 2	32 tons	Load 2	40 tons	Load 2	24 tons
Load 3	24 tons	Load 3	32 tons	Load 3	24 tons

a. Write a matrix for the amount of gravel in each load.

b. What are the dimensions of the matrix?

4-2 Skills Practice

Operations with Matrices

Perform the indicated operations. If the matrix does not exist, write *impossible*.

1. $[5 \ \ -4] + [4 \ \ 5]$

2. $\begin{bmatrix} 8 & 3 \\ -1 & -1 \end{bmatrix} - \begin{bmatrix} 0 & -7 \\ 6 & 2 \end{bmatrix}$

3. $[3 \ \ 1 \ \ 6] + \begin{bmatrix} 4 \\ -1 \\ 2 \end{bmatrix}$

4. $\begin{bmatrix} 5 & -1 & 2 \\ 1 & 8 & -6 \end{bmatrix} + \begin{bmatrix} 9 & 9 & 2 \\ 4 & 6 & 4 \end{bmatrix}$

5. $3[9 \ \ 4 \ \ -3]$

6. $[6 \ \ -3] - 4[4 \ \ 7]$

7. $-2\begin{bmatrix} -2 & 5 \\ 5 & 9 \end{bmatrix} + \begin{bmatrix} 1 & 1 \\ 1 & 1 \end{bmatrix}$

8. $3\begin{bmatrix} 8 \\ 0 \\ -3 \end{bmatrix} - 4\begin{bmatrix} 2 \\ 2 \\ 10 \end{bmatrix}$

9. $5\begin{bmatrix} -4 & 6 \\ 10 & 1 \\ -1 & 1 \end{bmatrix} + 2\begin{bmatrix} 6 & 5 \\ -3 & -2 \\ 1 & 0 \end{bmatrix}$

10. $3\begin{bmatrix} 3 & 1 & 3 \\ -4 & 7 & 5 \end{bmatrix} - 2\begin{bmatrix} 1 & -1 & 5 \\ 6 & 6 & -3 \end{bmatrix}$

Use matrices *A*, *B*, and *C* to find the following.

$A = \begin{bmatrix} 3 & 2 \\ 4 & 3 \end{bmatrix}$, $B = \begin{bmatrix} 2 & 2 \\ 1 & -2 \end{bmatrix}$, and $C = \begin{bmatrix} -3 & 4 \\ 3 & 1 \end{bmatrix}$.

11. $A + B$

12. $B - C$

13. $B - A$

14. $A + B + C$

15. $3B$

16. $-5C$

17. $A - 4C$

18. $2B + 3A$

NAME ______________________________ DATE ____________ PERIOD ________

4-2 Practice

Operations with Matrices

Perform the indicated operations. If the matrix does not exist, write *impossible*.

1. $\begin{bmatrix} 2 & -1 \\ 3 & 7 \\ 14 & -9 \end{bmatrix} + \begin{bmatrix} -6 & 9 \\ 7 & -11 \\ -8 & 17 \end{bmatrix}$

2. $\begin{bmatrix} 4 \\ -71 \\ 18 \end{bmatrix} - \begin{bmatrix} -67 \\ 45 \\ -24 \end{bmatrix}$

3. $-3\begin{bmatrix} -1 & 0 \\ 17 & -11 \end{bmatrix} + 4\begin{bmatrix} -3 & 16 \\ -21 & 12 \end{bmatrix}$

4. $7\begin{bmatrix} 2 & -1 & 8 \\ 4 & 7 & 9 \end{bmatrix} - 2\begin{bmatrix} -1 & 4 & -3 \\ 7 & 2 & -6 \end{bmatrix}$

5. $-2\begin{bmatrix} 1 \\ 2 \end{bmatrix} + 4\begin{bmatrix} 0 \\ 5 \end{bmatrix} - \begin{bmatrix} 10 \\ 18 \end{bmatrix}$

6. $\frac{3}{4}\begin{bmatrix} 8 & 12 \\ -16 & 20 \end{bmatrix} + \frac{2}{3}\begin{bmatrix} 27 & -9 \\ 54 & -18 \end{bmatrix}$

Use matrices $A = \begin{bmatrix} 4 & -1 & 0 \\ -3 & 6 & 2 \end{bmatrix}$, $B = \begin{bmatrix} -2 & 4 & 5 \\ 1 & 0 & 9 \end{bmatrix}$, and $C = \begin{bmatrix} 10 & -8 & 6 \\ -6 & -4 & 20 \end{bmatrix}$ to find the following.

7. $A - B$

8. $A - C$

9. $-3B$

10. $4B - A$

11. $-2B - 3C$

12. $A + 0.5C$

13. **ECONOMICS** Use the table that shows loans by an economic development board to women and men starting new businesses.

	Women		Men	
	Businesses	**Loan Amount ($)**	**Businesses**	**Loan Amount ($)**
2003	27	$567,000	36	$864,000
2004	41	$902,000	32	$672,000
2005	35	$777,000	28	$562,000

a. Write two matrices that represent the number of new businesses and loan amounts, one for women and one for men.

b. Find the sum of the numbers of new businesses and loan amounts for both men and women over the three-year period expressed as a matrix.

14. **PET NUTRITION** Use the table that gives nutritional information for two types of dog food. Find the difference in the percent of protein, fat, and fiber between Mix B and Mix A expressed as a matrix.

	% Protein	% Fat	% Fiber
Mix A	22	12	5
Mix B	24	8	8

NAME ______________________ DATE ____________ PERIOD ____________

4-3 Skills Practice

Multiplying Matrices

Determine whether each matrix product is defined. If so, state the dimensions of the product.

1. $A_{2 \times 5} \cdot B_{5 \times 1}$

2. $M_{1 \times 3} \cdot N_{3 \times 2}$

3. $B_{3 \times 2} \cdot A_{3 \times 2}$

4. $R_{4 \times 4} \cdot S_{4 \times 1}$

5. $X_{3 \times 3} \cdot Y_{3 \times 4}$

6. $A_{6 \times 4} \cdot B_{4 \times 5}$

Find each product, if possible.

7. $\begin{bmatrix} 3 & 2 \end{bmatrix} \cdot \begin{bmatrix} 2 \\ 1 \end{bmatrix}$

8. $\begin{bmatrix} 5 & 6 \\ 2 & 1 \end{bmatrix} \cdot \begin{bmatrix} 2 & -5 \\ 3 & 1 \end{bmatrix}$

9. $\begin{bmatrix} 1 & 3 \\ -1 & 1 \end{bmatrix} \cdot \begin{bmatrix} 3 \\ -2 \end{bmatrix}$

10. $\begin{bmatrix} 3 \\ -2 \end{bmatrix} \cdot \begin{bmatrix} 1 & 3 \\ -1 & 1 \end{bmatrix}$

11. $\begin{bmatrix} -3 & 4 \end{bmatrix} \cdot \begin{bmatrix} 0 & -1 \\ 2 & 2 \end{bmatrix}$

12. $\begin{bmatrix} -1 \\ 3 \end{bmatrix} \cdot \begin{bmatrix} 2 & -3 & -2 \end{bmatrix}$

13. $\begin{bmatrix} 5 \\ 6 \\ -3 \end{bmatrix} \cdot \begin{bmatrix} 4 \\ 8 \end{bmatrix}$

14. $\begin{bmatrix} 2 & -2 \\ 4 & 5 \\ -3 & 1 \end{bmatrix} \cdot \begin{bmatrix} 0 & 3 \\ 3 & 0 \end{bmatrix}$

15. $\begin{bmatrix} -4 & 4 \\ -2 & 1 \\ 2 & 3 \end{bmatrix} \cdot \begin{bmatrix} 3 & -3 \\ 0 & 2 \end{bmatrix}$

16. $\begin{bmatrix} 0 & 1 & 1 \\ 1 & 1 & 0 \end{bmatrix} \cdot \begin{bmatrix} 2 \\ 2 \\ 2 \end{bmatrix}$

Use $A = \begin{bmatrix} 2 & 1 \\ 2 & 1 \end{bmatrix}$, $B = \begin{bmatrix} -3 & 2 \\ 5 & 1 \end{bmatrix}$, $C = \begin{bmatrix} 3 & -1 \\ 1 & 0 \end{bmatrix}$, and $c = 2$ to determine whether the following equations are true for the given matrices.

17. $c(AC) = A(cC)$

18. $AB = BA$

19. $B(A + C) = AB + BC$

20. $c(A - B) = cA - cB$

NAME ______________________ DATE ____________ PERIOD ________

4-3 Practice

Multiplying Matrices

Determine whether each matrix product is defined. If so, state the dimensions of the product.

1. $A_{7 \times 4} \cdot B_{4 \times 3}$
2. $A_{3 \times 5} \cdot M_{5 \times 8}$
3. $M_{2 \times 1} \cdot A_{1 \times 6}$
4. $M_{3 \times 2} \cdot A_{3 \times 2}$
5. $P_{1 \times 9} \cdot Q_{9 \times 1}$
6. $P_{9 \times 1} \cdot Q_{1 \times 9}$

Find each product, if possible.

7. $\begin{bmatrix} 2 & 4 \\ 3 & -1 \end{bmatrix} \cdot \begin{bmatrix} 3 & -2 & 7 \\ 6 & 0 & -5 \end{bmatrix}$

8. $\begin{bmatrix} 2 & 4 \\ 7 & -1 \end{bmatrix} \cdot \begin{bmatrix} -3 & 0 \\ 2 & 5 \end{bmatrix}$

9. $\begin{bmatrix} -3 & 0 \\ 2 & 5 \end{bmatrix} \cdot \begin{bmatrix} 2 & 4 \\ 7 & -1 \end{bmatrix}$

10. $\begin{bmatrix} 3 & -2 & 7 \\ 6 & 0 & -5 \end{bmatrix} \cdot \begin{bmatrix} 3 & -2 & 7 \\ 6 & 0 & -5 \end{bmatrix}$

11. $[4 \ \ 0 \ \ 2] \cdot \begin{bmatrix} 1 \\ 3 \\ -1 \end{bmatrix}$

12. $\begin{bmatrix} 1 \\ 3 \\ -1 \end{bmatrix} \cdot [4 \ \ 0 \ \ 2]$

13. $\begin{bmatrix} -6 & 2 \\ 3 & -1 \end{bmatrix} \cdot \begin{bmatrix} 5 & 0 \\ 0 & 5 \end{bmatrix}$

14. $[-15 \ \ -9] \cdot \begin{bmatrix} 6 & 11 \\ 23 & -10 \end{bmatrix}$

Use $A = \begin{bmatrix} 1 & 3 \\ 3 & 1 \end{bmatrix}$, $B = \begin{bmatrix} 4 & 0 \\ -2 & -1 \end{bmatrix}$, $C = \begin{bmatrix} -1 & 0 \\ 0 & -1 \end{bmatrix}$, and $c = 3$ to determine whether the following equations are true for the given matrices.

15. $AC = CA$
16. $A(B + C) = BA + CA$
17. $(AB)c = c(AB)$
18. $(A + C)B = B(A + C)$

19. **RENTALS** For their one-week vacation, the Montoyas can rent a 2-bedroom condominium for $1796, a 3-bedroom condominium for $2165, or a 4-bedroom condominium for $2538. The table shows the number of units in each of three complexes.

	2-Bedroom	3-Bedroom	4-Bedroom
Sun Haven	36	24	22
Surfside	29	32	42
Seabreeze	18	22	18

a. Write a matrix that represents the number of each type of unit available at each complex and a matrix that represents the weekly charge for each type of unit.

b. If all of the units in the three complexes are rented for the week at the rates given the Montoyas, express the income of each of the three complexes as a matrix.

c. What is the total income of all three complexes for the week?

NAME ______________________________ DATE ____________ PERIOD ________

4-4 Skills Practice

Transformations with Matrices

1. Triangle ABC with vertices $A(2, 3)$, $B(0, 4)$, and $C(-3, -3)$ is translated 3 units right and 1 unit down.

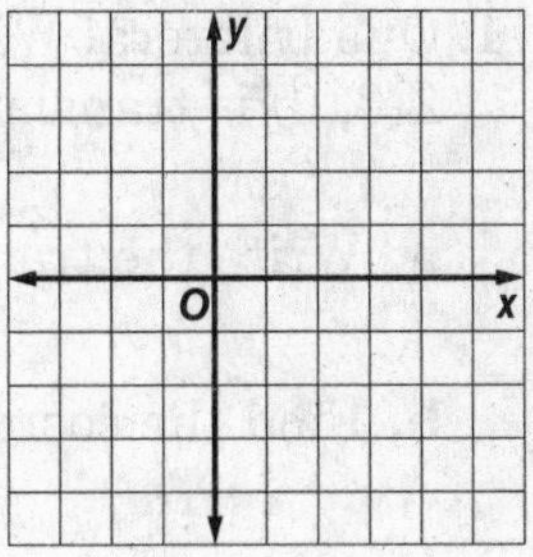

a. Write the translation matrix.

b. Find the coordinates of $\triangle A'B'C'$.

c. Graph the preimage and the image.

2. The vertices of $\triangle RST$ are $R(-3, 1)$, $S(2, -1)$, and $T(1, 3)$. The triangle is dilated so that its perimeter is twice the original perimeter.

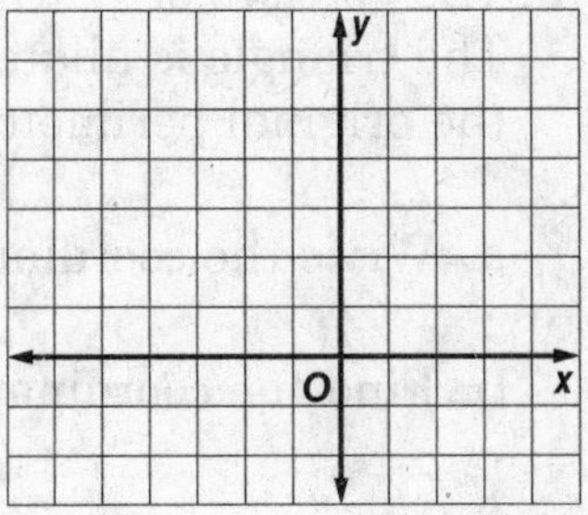

a. Write the coordinates of $\triangle RST$ in a vertex matrix.

b. Find the coordinates of the image $\triangle R'S'T'$.

c. Graph $\triangle RST$ and $\triangle R'S'T'$.

3. The vertices of $\triangle DEF$ are $D(4, 0)$, $E(0, -1)$, and $F(2, 3)$. The triangle is reflected over the x-axis.

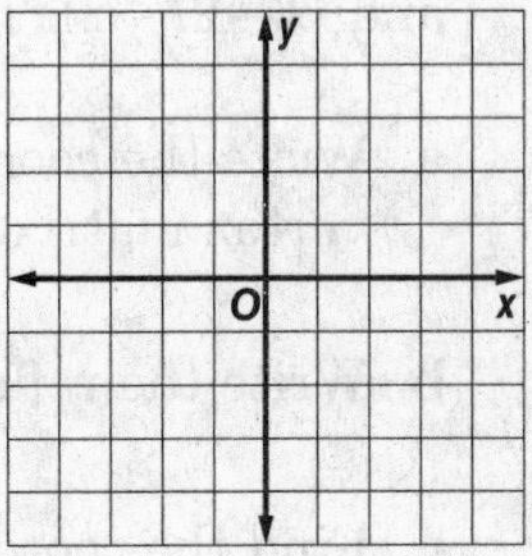

a. Write the coordinates of $\triangle DEF$ in a vertex matrix.

b. Write the reflection matrix for this situation.

c. Find the coordinates of $\triangle D'E'F'$.

d. Graph $\triangle DEF$ and $\triangle D'E'F'$.

4. Triangle XYZ with vertices $X(1, -3)$, $Y(-4, 1)$, and $Z(-2, 5)$ is rotated 180° counterclockwise about the origin.

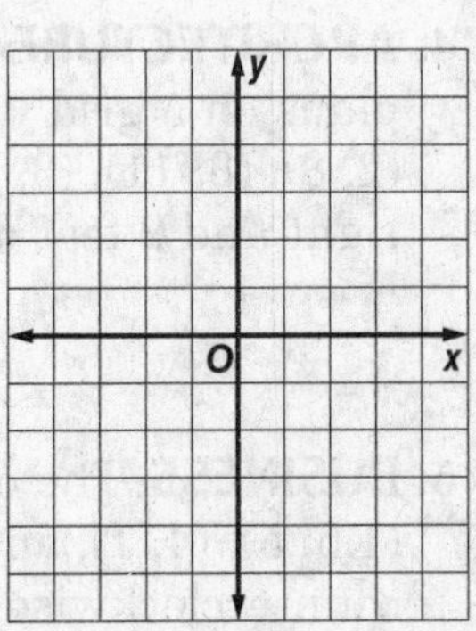

a. Write the coordinates of $\triangle XYZ$ in a vertex matrix.

b. Write the rotation matrix for this situation.

c. Find the coordinates of $\triangle X'Y'Z'$.

d. Graph the preimage and the image.

4-4 Practice

Transformations with Matrices

1. Quadrilateral *WXYZ* with vertices $W(-3, 2)$, $X(-2, 4)$, $Y(4, 1)$, and $Z(3, 0)$ is translated 1 unit left and 3 units down.

a. Write the translation matrix.

b. Find the coordinates of quadrilateral $W'X'Y'Z'$.

c. Graph the preimage and the image.

2. The vertices of $\triangle RST$ are $R(6, 2)$, $S(3, -3)$, and $T(-2, 5)$. The triangle is dilated so that its perimeter is one half the original perimeter.

a. Write the coordinates of $\triangle RST$ in a vertex matrix.

b. Find the coordinates of the image $\triangle R'S'T'$.

c. Graph $\triangle RST$ and $\triangle R'S'T'$.

3. The vertices of quadrilateral *ABCD* are $A(-3, 2)$, $B(0, 3)$, $C(4, -4)$, and $D(-2, -2)$. The quadrilateral is reflected in the *y*-axis.

a. Write the coordinates of *ABCD* in a vertex matrix.

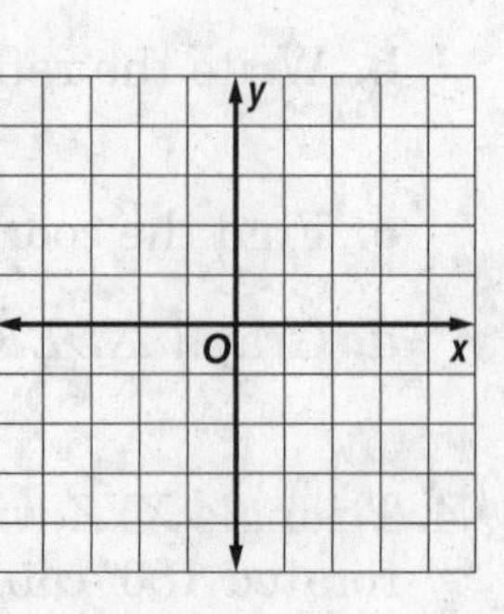

b. Write the reflection matrix for this situation.

c. Find the coordinates of $A'B'C'D'$.

d. Graph *ABCD* and $A'B'C'D'$.

4. ARCHITECTURE Using architectural design software, the Bradleys plot their kitchen plans on a grid with each unit representing 1 foot. They place the corners of an island at (2, 8), (8, 11), (3, 5), and (9, 8). If the Bradleys wish to move the island 1.5 feet to the right and 2 feet down, what will the new coordinates of its corners be?

5. BUSINESS The design of a business logo calls for locating the vertices of a triangle at (1.5, 5), (4, 1), and (1, 0) on a grid. If design changes require rotating the triangle 90° counterclockwise, what will the new coordinates of the vertices be?

4-5 Skills Practice

Determinants and Cramer's Rule

Evaluate each determinant.

1. $\begin{vmatrix} 5 & 2 \\ 1 & 3 \end{vmatrix}$

2. $\begin{vmatrix} 10 & 9 \\ 5 & 8 \end{vmatrix}$

3. $\begin{vmatrix} 1 & 6 \\ 1 & 7 \end{vmatrix}$

4. $\begin{vmatrix} 2 & 5 \\ 3 & 1 \end{vmatrix}$

5. $\begin{vmatrix} 0 & 9 \\ 5 & 8 \end{vmatrix}$

6. $\begin{vmatrix} 3 & 12 \\ 2 & 8 \end{vmatrix}$

7. $\begin{vmatrix} -5 & 2 \\ 8 & -6 \end{vmatrix}$

8. $\begin{vmatrix} -3 & 1 \\ 8 & -7 \end{vmatrix}$

9. $\begin{vmatrix} 9 & -2 \\ -4 & 1 \end{vmatrix}$

10. $\begin{vmatrix} 1 & -5 \\ 1 & 6 \end{vmatrix}$

11. $\begin{vmatrix} 1 & -3 \\ -3 & 4 \end{vmatrix}$

12. $\begin{vmatrix} -12 & 4 \\ 1 & 4 \end{vmatrix}$

13. $\begin{vmatrix} 3 & -5 \\ 6 & -11 \end{vmatrix}$

14. $\begin{vmatrix} -1 & -3 \\ 5 & -2 \end{vmatrix}$

15. $\begin{vmatrix} -1 & -14 \\ 5 & 2 \end{vmatrix}$

16. $\begin{vmatrix} -1 & 2 \\ 0 & 4 \end{vmatrix}$

17. $\begin{vmatrix} 2 & 2 \\ -1 & 4 \end{vmatrix}$

18. $\begin{vmatrix} -1 & 6 \\ 2 & 5 \end{vmatrix}$

Evaluate each determinant using diagonals.

19. $\begin{vmatrix} 2 & -1 & 1 \\ 3 & 2 & -1 \\ 2 & 3 & -2 \end{vmatrix}$

20. $\begin{vmatrix} 6 & -1 & 1 \\ 5 & 2 & -1 \\ 1 & 3 & -2 \end{vmatrix}$

21. $\begin{vmatrix} 2 & 6 & 1 \\ 3 & 5 & -1 \\ 2 & 1 & -2 \end{vmatrix}$

22. $\begin{vmatrix} 2 & -1 & 6 \\ 3 & 2 & 5 \\ 2 & 3 & 1 \end{vmatrix}$

23. $\begin{vmatrix} 3 & -1 & 2 \\ 1 & 0 & 4 \\ 3 & -2 & 0 \end{vmatrix}$

24. $\begin{vmatrix} 3 & 2 & 2 \\ 1 & -1 & 4 \\ 3 & -1 & 0 \end{vmatrix}$

4-5 Practice

Determinants and Cramer's Rule

Evaluate each determinant.

1. $\begin{vmatrix} 1 & 6 \\ 2 & 7 \end{vmatrix}$

2. $\begin{vmatrix} 9 & 6 \\ 3 & 2 \end{vmatrix}$

3. $\begin{vmatrix} 4 & 1 \\ -2 & -5 \end{vmatrix}$

4. $\begin{vmatrix} -14 & -3 \\ 2 & -2 \end{vmatrix}$

5. $\begin{vmatrix} 4 & -3 \\ -12 & 4 \end{vmatrix}$

6. $\begin{vmatrix} 2 & -5 \\ 5 & -11 \end{vmatrix}$

7. $\begin{vmatrix} 3 & -4 \\ 3.75 & 5 \end{vmatrix}$

8. $\begin{vmatrix} 2 & -1 \\ 3 & -9.5 \end{vmatrix}$

9. $\begin{vmatrix} 0.5 & -0.7 \\ 0.4 & -0.3 \end{vmatrix}$

Evaluate each determinant using expansion by diagonals.

10. $\begin{vmatrix} -2 & 3 & 1 \\ 0 & 4 & -3 \\ 2 & 5 & -1 \end{vmatrix}$

11. $\begin{vmatrix} 2 & -4 & 1 \\ 3 & 0 & 9 \\ -1 & 5 & 7 \end{vmatrix}$

12. $\begin{vmatrix} 2 & 1 & 1 \\ 1 & -1 & -2 \\ 1 & 1 & -1 \end{vmatrix}$

13. $\begin{vmatrix} 0 & -4 & 0 \\ 2 & -1 & 1 \\ 3 & -2 & 5 \end{vmatrix}$

14. $\begin{vmatrix} 2 & 7 & -6 \\ 8 & 4 & 0 \\ 1 & -1 & 3 \end{vmatrix}$

15. $\begin{vmatrix} -12 & 0 & 3 \\ 7 & 5 & -1 \\ 4 & 2 & -6 \end{vmatrix}$

Use Cramer's Rule to solve each system of equation.

16. $4x - 2y = -6$
$3x + y = 18$

17. $5x + 4y = 10$
$-3x - 2y = -8$

18. $-2x - 3y = -14$
$4x - y = 0$

19. $6x + 6y = 9$
$4x - 4y = -42$

20. $5x - 6 = 3y$
$5y = 54 + 3x$

21. $\frac{x}{2} + \frac{y}{4} = 2$
$\frac{x}{4} - \frac{y}{6} = -6$

25. **GEOMETRY** Find the area of a triangle whose vertices have coordinates (3, 5), (6, –5), and (–4, 10).

26. **LAND MANAGEMENT** A fish and wildlife management organization uses a GIS (geographic information system) to store and analyze data for the parcels of land it manages. All of the parcels are mapped on a grid in which 1 unit represents 1 acre. If the coordinates of the corners of a parcel are (–8, 10), (6, 17), and (2, –4), how many acres is the parcel?

NAME ______________________ DATE ____________ PERIOD ____________

4-6 Skills Practice

Inverse Matrices and Systems of Equations

Determine whether the matrices in each pair are inverses.

1. $X = \begin{bmatrix} 1 & 0 \\ 1 & 1 \end{bmatrix}, Y = \begin{bmatrix} -1 & 0 \\ 1 & 1 \end{bmatrix}$

2. $P = \begin{bmatrix} 2 & 3 \\ 1 & 1 \end{bmatrix}, Q = \begin{bmatrix} -1 & 3 \\ 1 & -2 \end{bmatrix}$

3. $M = \begin{bmatrix} -1 & 0 \\ 0 & 3 \end{bmatrix}, N = \begin{bmatrix} -1 & 0 \\ 0 & -3 \end{bmatrix}$

4. $A = \begin{bmatrix} -2 & 5 \\ -1 & 2 \end{bmatrix}, B = \begin{bmatrix} 2 & -5 \\ 1 & -2 \end{bmatrix}$

5. $V = \begin{bmatrix} 0 & 7 \\ -7 & 0 \end{bmatrix}, W = \begin{bmatrix} 0 & -\frac{1}{7} \\ \frac{1}{7} & 0 \end{bmatrix}$

6. $X = \begin{bmatrix} -1 & 4 \\ 1 & 2 \end{bmatrix}, Y = \begin{bmatrix} -\frac{1}{3} & \frac{2}{3} \\ \frac{1}{6} & \frac{1}{6} \end{bmatrix}$

7. $G = \begin{bmatrix} 4 & -3 \\ 1 & 2 \end{bmatrix}, H = \begin{bmatrix} \frac{2}{11} & \frac{3}{11} \\ -\frac{1}{11} & \frac{4}{11} \end{bmatrix}$

8. $D = \begin{bmatrix} -4 & -4 \\ -4 & 4 \end{bmatrix}, E = \begin{bmatrix} -0.125 & -0.125 \\ -0.125 & -0.125 \end{bmatrix}$

Find the inverse of each matrix, if it exists.

9. $\begin{bmatrix} 0 & 2 \\ 4 & 0 \end{bmatrix}$

10. $\begin{bmatrix} 1 & 1 \\ 3 & 2 \end{bmatrix}$

11. $\begin{bmatrix} 9 & 3 \\ 6 & 2 \end{bmatrix}$

12. $\begin{bmatrix} -2 & -4 \\ 6 & 0 \end{bmatrix}$

13. $\begin{bmatrix} 1 & -1 \\ 3 & 3 \end{bmatrix}$

14. $\begin{bmatrix} 3 & 6 \\ -1 & -2 \end{bmatrix}$

Use a matrix equation to solve each system of equations.

15. $p - 3q = 6$
 $2p + 3q = -6$

16. $-x - 3y = 2$
 $-4x - 5y = 1$

17. $2m + 2n = -8$
 $6m + 4n = -18$

18. $-3a + b = -9$
 $5a - 2b = 14$

NAME ______________________ DATE ____________ PERIOD ____________

4-6 Practice

Inverse Matrices and Systems of Equations

Determine whether each pair of matrices are inverses.

1. $M = \begin{bmatrix} 2 & 1 \\ 3 & 2 \end{bmatrix}, N = \begin{bmatrix} -2 & 1 \\ 3 & -2 \end{bmatrix}$

2. $X = \begin{bmatrix} -3 & 2 \\ 5 & -3 \end{bmatrix}, Y = \begin{bmatrix} 3 & 2 \\ 5 & 3 \end{bmatrix}$

3. $A = \begin{bmatrix} 3 & 1 \\ -4 & 2 \end{bmatrix}, B = \begin{bmatrix} \frac{1}{5} & -\frac{1}{10} \\ \frac{2}{5} & \frac{3}{10} \end{bmatrix}$

4. $P = \begin{bmatrix} 6 & -2 \\ -2 & 3 \end{bmatrix}, Q = \begin{bmatrix} \frac{3}{14} & \frac{1}{7} \\ \frac{1}{7} & \frac{3}{7} \end{bmatrix}$

Determine whether each statement is *true* or *false*.

5. All square matrices have multiplicative inverses.

6. All square matrices have multiplicative identities.

Find the inverse of each matrix, if it exists.

7. $\begin{bmatrix} 4 & 5 \\ -4 & -3 \end{bmatrix}$

8. $\begin{bmatrix} 2 & 0 \\ 3 & 5 \end{bmatrix}$

9. $\begin{bmatrix} -1 & 3 \\ 4 & -7 \end{bmatrix}$

10. $\begin{bmatrix} 2 & 5 \\ -1 & 3 \end{bmatrix}$

11. $\begin{bmatrix} 2 & -5 \\ 3 & 1 \end{bmatrix}$

12. $\begin{bmatrix} 4 & 6 \\ 6 & 9 \end{bmatrix}$

13. **GEOMETRY** Use the figure at the right.

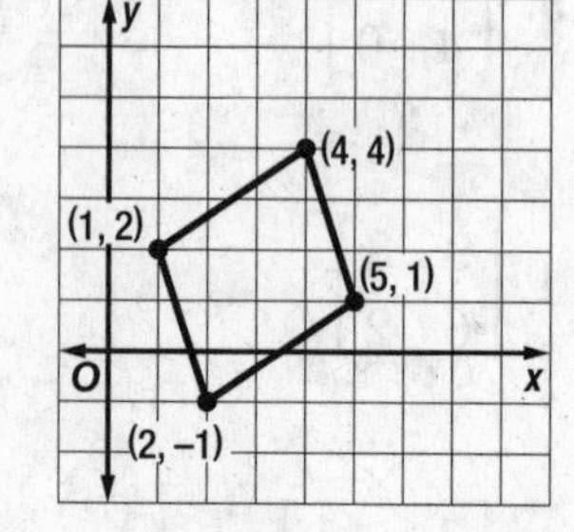

a. Write the vertex matrix A for the rectangle.

b. Use matrix multiplication to find BA if $B = \begin{bmatrix} 1.5 & 0 \\ 0 & 1.5 \end{bmatrix}$.

c. Graph the vertices of the transformed quadrilateral on the previous graph. Describe the transformation.

d. Make a conjecture about what transformation B^{-1} describes on a coordinate plane.

14. **CODES** Use the alphabet table below and the inverse of coding matrix $C = \begin{bmatrix} 1 & 2 \\ 2 & 1 \end{bmatrix}$ to decode this message:
19 | 14 | 11 | 13 | 11 | 22 | 55 | 65 | 57 | 60 | 2 | 1 | 52 | 47 | 33 | 51 | 56 | 55.

CODE													
A	1	B	2	C	3	D	4	E	5	F	6	G	7
H	8	I	9	J	10	K	11	L	12	M	13	N	14
O	15	P	16	Q	17	R	18	S	19	T	20	U	21
V	22	W	23	X	24	Y	25	Z	26	–	0		

NAME ________________________________ DATE ______________ PERIOD ____________

5-1 Skills Practice

Graphing Quadratic Functions

Complete parts a–c for each quadratic function.

a. Find the y-intercept, the equation of the axis of symmetry, and the x-coordinate of the vertex.

b. Make a table of values that includes the vertex.

c. Use this information to graph the function.

1. $f(x) = -2x^2$

2. $f(x) = x^2 - 4x + 4$

3. $f(x) = x^2 - 6x + 8$

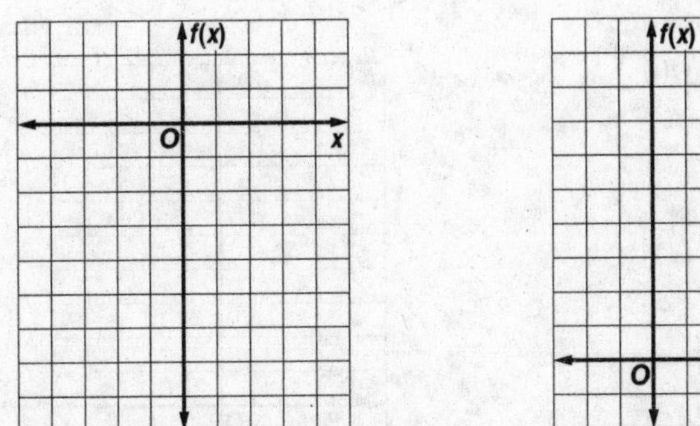

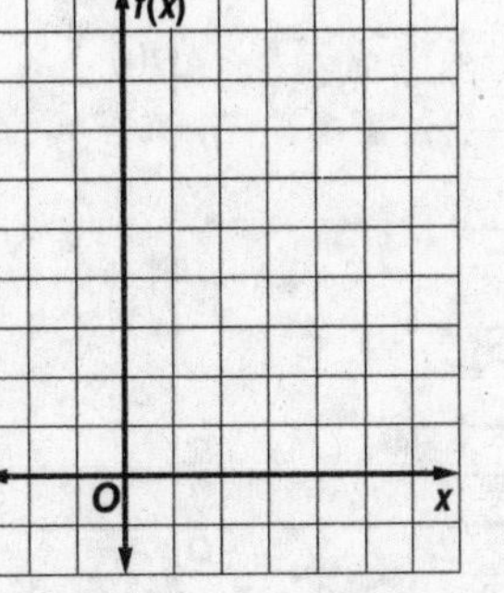

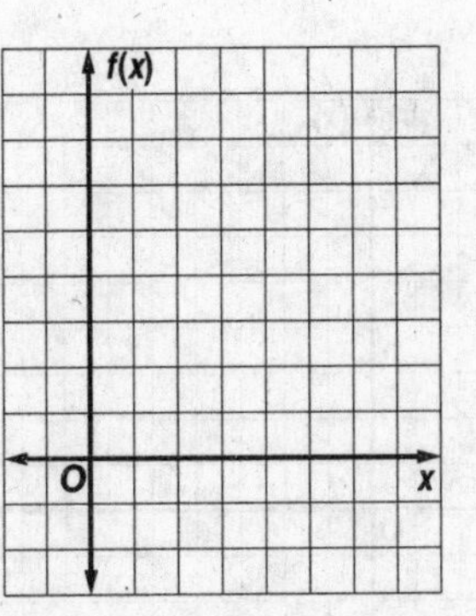

Determine whether each function has a maximum or a minimum value, and find that value. Then state the domain and range of the function.

4. $f(x) = 6x^2$

5. $f(x) = -8x^2$

6. $f(x) = x^2 + 2x$

7. $f(x) = -2x^2 + 4x - 3$

8. $f(x) = 3x^2 + 12x + 3$

9. $f(x) = 2x^2 + 4x + 1$

10. $f(x) = 3x^2$

11. $f(x) = x^2 + 1$

12. $f(x) = -x^2 + 6x - 15$

13. $f(x) = 2x^2 - 11$

14. $f(x) = x^2 - 10x + 5$

15. $f(x) = -2x^2 + 8x + 7$

5-1 Practice

Graphing Quadratic Functions

Complete parts a–c for each quadratic function.

a. Find the *y*-intercept, the equation of the axis of symmetry, and the *x*-coordinate of the vertex.

b. Make a table of values that includes the vertex.

c. Use this information to graph the function.

1. $f(x) = x^2 - 8x + 15$ **2.** $f(x) = -x^2 - 4x + 12$ **3.** $f(x) = 2x^2 - 2x + 1$

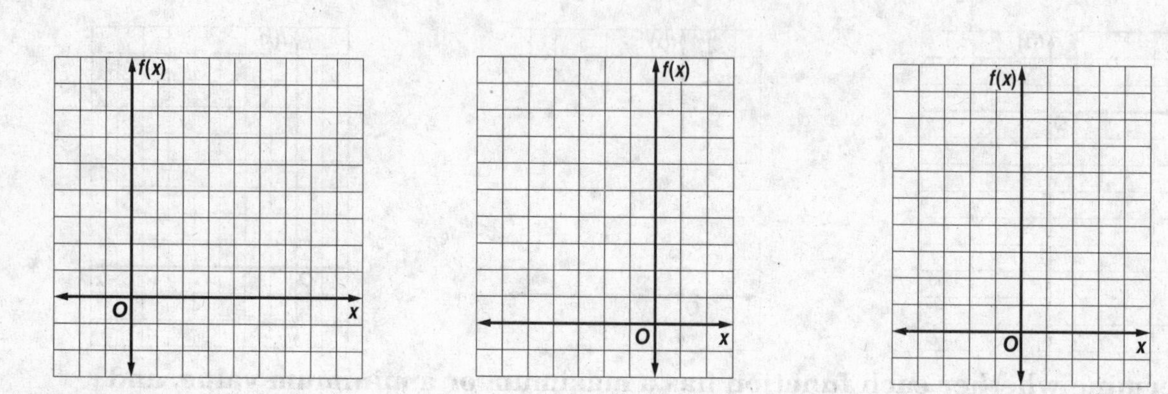

Determine whether each function has a *maximum* or *minimum* value, and find that value. Then state the domain and range of the function.

4. $f(x) = x^2 + 2x - 8$ **5.** $f(x) = x^2 - 6x + 14$ **6.** $v(x) = -x^2 + 14x - 57$

7. $f(x) = 2x^2 + 4x - 6$ **8.** $f(x) = -x^2 + 4x - 1$ **9.** $f(x) = -\frac{2}{3}x^2 + 8x - 24$

10. GRAVITATION From 4 feet above a swimming pool, Susan throws a ball upward with a velocity of 32 feet per second. The height $h(t)$ of the ball t seconds after Susan throws it is given by $h(t) = -16t^2 + 32t + 4$. For $t \geq 0$, find the maximum height reached by the ball and the time that this height is reached.

11. HEALTH CLUBS Last year, the SportsTime Athletic Club charged $20 to participate in an aerobics class. Seventy people attended the classes. The club wants to increase the class price this year. They expect to lose one customer for each $1 increase in the price.

a. What price should the club charge to maximize the income from the aerobics classes?

b. What is the maximum income the SportsTime Athletic Club can expect to make?

NAME ______________________ DATE __________ PERIOD __________

5-2 Skills Practice

Solving Quadratic Equations By Graphing

Use the related graph of each equation to determine its solutions.

1. $x^2 + 2x - 3 = 0$

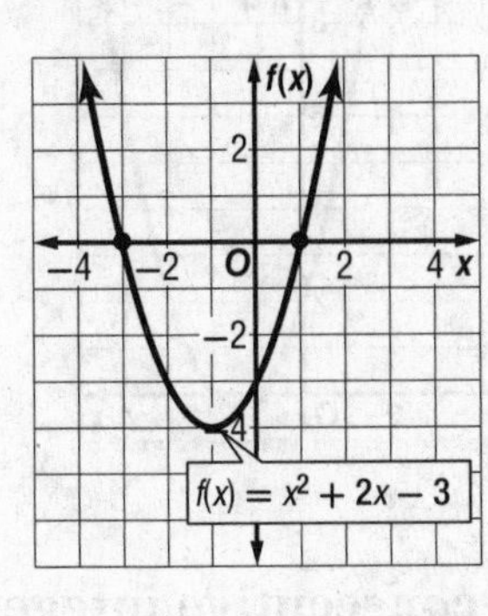

2. $-x^2 - 6x - 9 = 0$

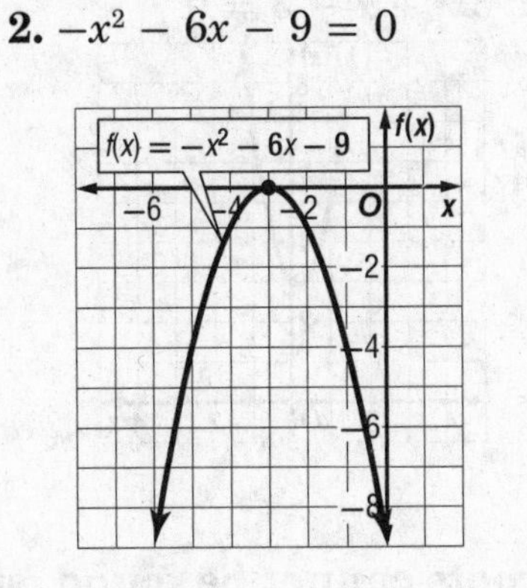

3. $3x^2 + 4x + 3 = 0$

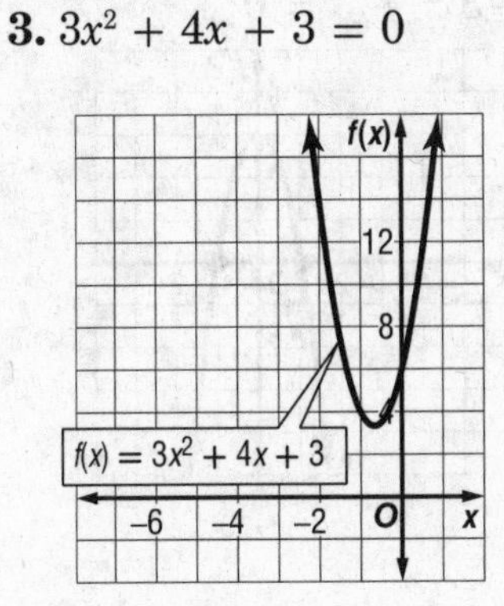

Solve each equation. If exact roots cannot be found, state the consecutive integers between which the roots are located.

4. $x^2 - 6x + 5 = 0$

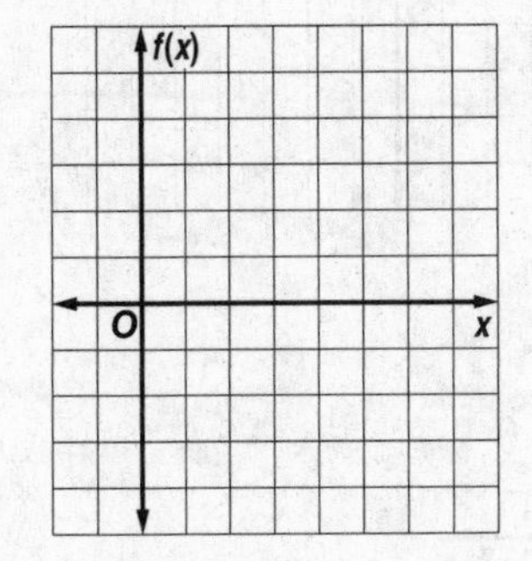

5. $-x^2 + 2x - 4 = 0$

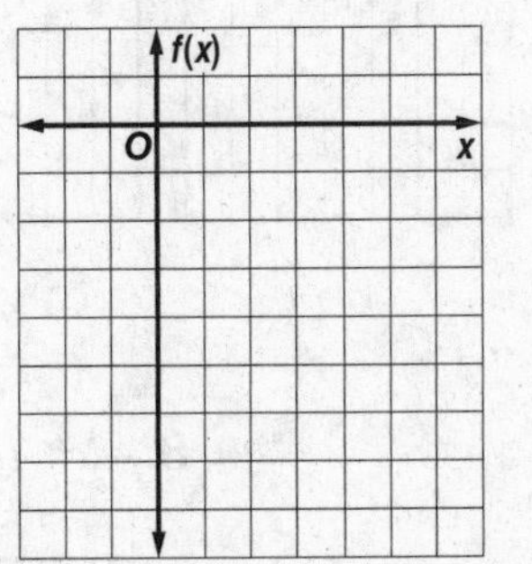

6. $x^2 - 6x + 4 = 0$

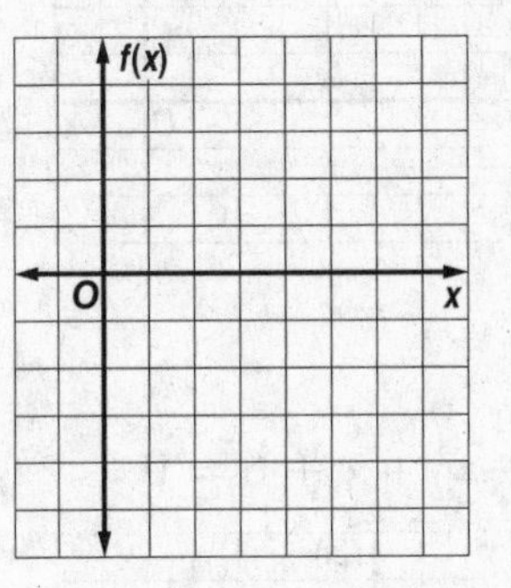

7. $-x^2 - 4x = 0$

f(x)

O x

8. $-x^2 + 36 = 0$

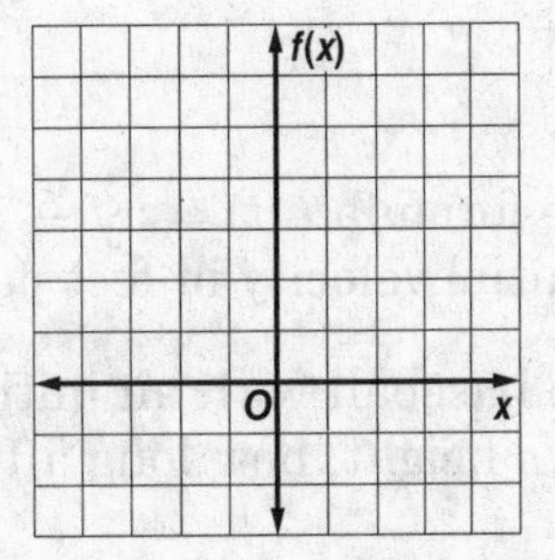

NAME ______________________ DATE ____________ PERIOD ________

5-2 Practice

Solving Quadratic Equations By Graphing

Use the related graph of each equation to determine its solutions.

1. $-3x^2 + 3 = 0$

2. $3x^2 + x + 3 = 0$

3. $x^2 - 3x + 2 = 0$

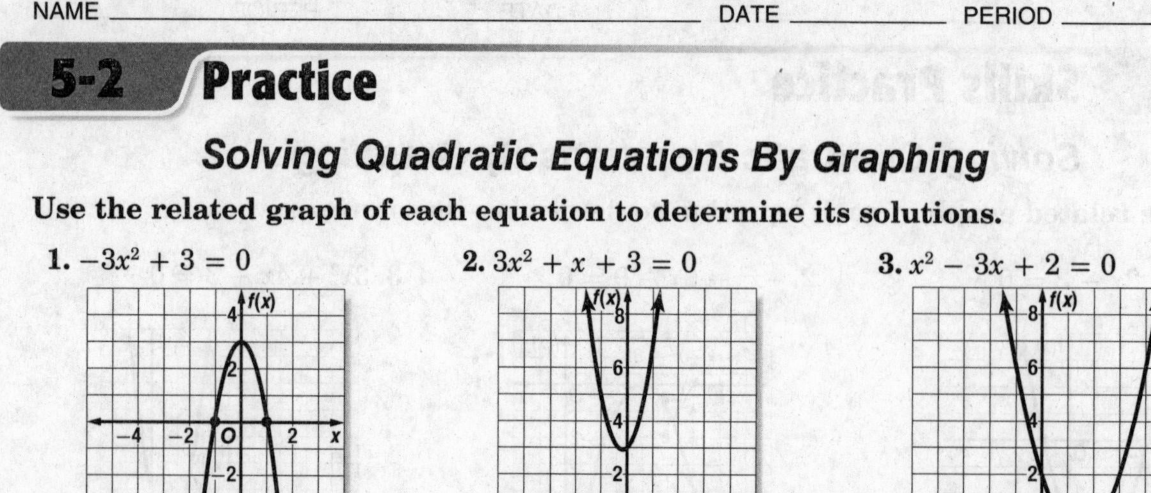

Solve each equation. If exact roots cannot be found, state the consecutive integers between which the roots are located.

4. $-2x^2 - 6x + 5 = 0$

5. $x^2 + 10x + 24 = 0$

6. $2x^2 - x - 6 = 0$

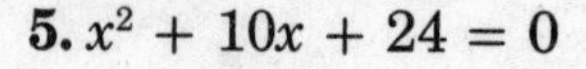

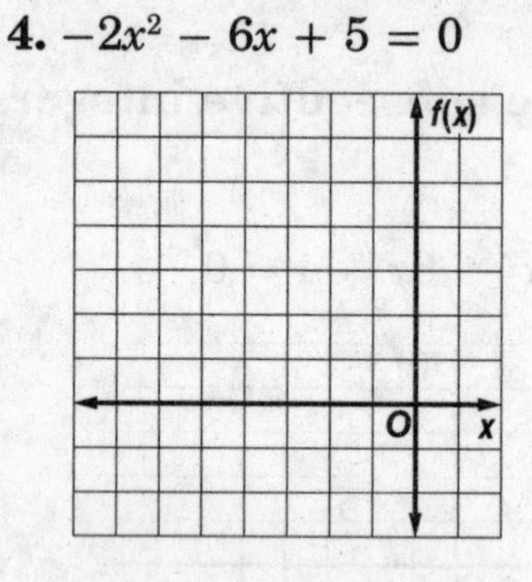

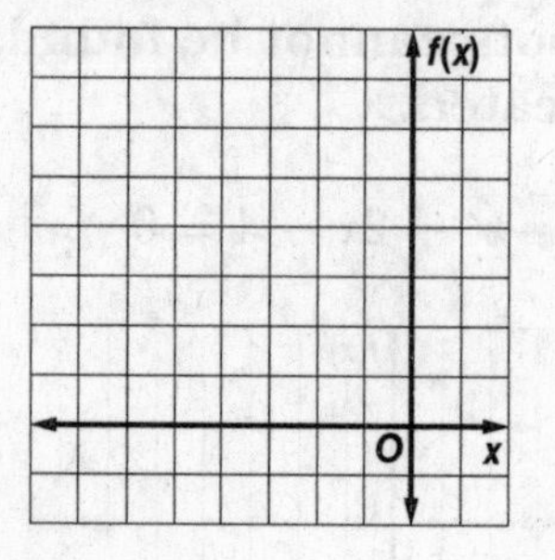

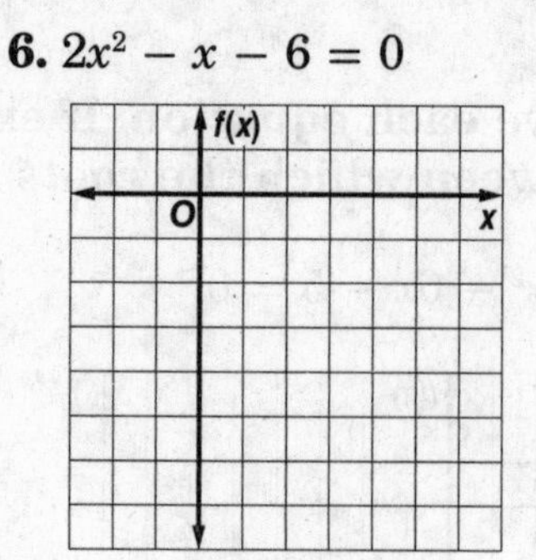

7. $-x^2 + x + 6 = 0$

8. $-x^2 + 5x - 8 = 0$

9. GRAVITY Use the formula $h(t) = v_0t - 16t^2$, where $h(t)$ is the height of an object in feet, v_0 is the object's initial velocity in feet per second, and t is the time in seconds.

a. Marta throws a baseball with an initial upward velocity of 60 feet per second. Ignoring Marta's height, how long after she releases the ball will it hit the ground?

b. A volcanic eruption blasts a boulder upward with an initial velocity of 240 feet per second. How long will it take the boulder to hit the ground if it lands at the same elevation from which it was ejected?

5-3 Skills Practice

Solving Quadratic Equations by Factoring

Write a quadratic equation in standard form with the given root(s).

1. 1, 4

2. 6, −9

3. −2, −5

4. 0, 7

5. $-\frac{1}{3}, -3$

6. $-\frac{1}{2}, \frac{3}{4}$

Factor each polynomial.

7. $m^2 + 7m - 18$

8. $2x^2 - 3x - 5$

9. $4z^2 + 4z - 15$

10. $4p^2 + 4p - 24$

11. $3y^2 + 21y + 36$

12. $c^2 - 100$

Solve each equation by factoring.

13. $x^2 = 64$

14. $x^2 - 100 = 0$

15. $x^2 - 3x + 2 = 0$

16. $x^2 - 4x + 3 = 0$

17. $x^2 + 2x - 3 = 0$

18. $x^2 - 3x - 10 = 0$

19. $x^2 - 6x + 5 = 0$

20. $x^2 - 9x = 0$

21. $x^2 - 4x = 21$

22. $2x^2 + 5x - 3 = 0$

23. $4x^2 + 5x - 6 = 0$

24. $3x^2 - 13x - 10 = 0$

25. NUMBER THEORY Find two consecutive integers whose product is 272.

NAME ______________________ DATE ____________ PERIOD ____________

5-3 Practice

Solving Quadratic Equations by Factoring

Write a quadratic equation in standard form with the given root(s).

1. 7, 2

2. 0, 3

3. –5, 8

4. –7, –8

5. –6, –3

6. 3, –4

7. $1, \frac{1}{2}$

8. $\frac{1}{3}, 2$

9. $0, -\frac{7}{2}$

Factor each polynomial.

10. $r^3 + 3r^2 - 54r$

11. $8a^2 + 2a - 6$

12. $c^2 - 49$

13. $x^3 + 8$

14. $16r^2 - 169$

15. $b^4 - 81$

Solve each equation by factoring.

16. $x^2 - 4x - 12 = 0$

17. $x^2 - 16x + 64 = 0$

18. $x^2 - 6x + 8 = 0$

19. $x^2 + 3x + 2 = 0$

20. $x^2 - 4x = 0$

21. $7x^2 = 4x$

22. $10x^2 = 9x$

23. $x^2 = 2x + 99$

24. $x^2 + 12x = -36$

25. $5x^2 - 35x + 60 = 0$

26. $36x^2 = 25$

27. $2x^2 - 8x - 90 = 0$

28. NUMBER THEORY Find two consecutive even positive integers whose product is 624.

29. NUMBER THEORY Find two consecutive odd positive integers whose product is 323.

30. GEOMETRY The length of a rectangle is 2 feet more than its width. Find the dimensions of the rectangle if its area is 63 square feet.

31. PHOTOGRAPHY The length and width of a 6-inch by 8-inch photograph are reduced by the same amount to make a new photograph whose area is half that of the original. By how many inches will the dimensions of the photograph have to be reduced?

NAME ______________________ DATE ____________ PERIOD ____________

5-4 Skills Practice

Complex Numbers

Simplify.

1. $\sqrt{99}$

2. $\sqrt{\frac{27}{49}}$

3. $\sqrt{52x^3y^5}$

4. $\sqrt{-108x^7}$

5. $\sqrt{-81x^6}$

6. $\sqrt{-23} \cdot \sqrt{-46}$

7. $(3\boldsymbol{i})(-2\boldsymbol{i})(5\boldsymbol{i})$

8. $\boldsymbol{i}^{11}$

9. $\boldsymbol{i}^{65}$

10. $(7 - 8\boldsymbol{i}) + (-12 - 4\boldsymbol{i})$

11. $(-3 + 5\boldsymbol{i}) + (18 - 7\boldsymbol{i})$

12. $(10 - 4\boldsymbol{i}) - (7 + 3\boldsymbol{i})$

13. $(7 - 6\boldsymbol{i})(2 - 3\boldsymbol{i})$

14. $(3 + 4\boldsymbol{i})(3 - 4\boldsymbol{i})$

15. $\frac{8 - 6\boldsymbol{i}}{3\boldsymbol{i}}$

16. $\frac{3\boldsymbol{i}}{4 + 2\boldsymbol{i}}$

Solve each equation.

17. $3x^2 + 3 = 0$

18. $5x^2 + 125 = 0$

19. $4x^2 + 20 = 0$

20. $-x^2 - 16 = 0$

21. $x^2 + 18 = 0$

22. $8x^2 + 96 = 0$

Find the values of ℓ and m that make each equation true.

23. $20 - 12\boldsymbol{i} = 5\ell + (4m)\boldsymbol{i}$

24. $\ell - 16\boldsymbol{i} = 3 - (2m)\boldsymbol{i}$

25. $(4 + \ell) + (2m)\boldsymbol{i} = 9 + 14\boldsymbol{i}$

26. $(3 - m) + (7\ell - 14)\boldsymbol{i} = 1 + 7\boldsymbol{i}$

NAME ______________________ DATE ____________ PERIOD ________

5-4 Practice

Complex Numbers

Simplify.

1. $\sqrt{-36}$

2. $\sqrt{-8} \cdot \sqrt{-32}$

3. $\sqrt{-15} \cdot \sqrt{-25}$

4. $(-3i)(4i)(-5i)$

5. $(7i)^2(6i)$

6. i^{42}

7. i^{55}

8. i^{89}

9. $(5 - 2i) + (-13 - 8i)$

10. $(7 - 6i) + (9 + 11i)$

11. $(-12 + 48i) + (15 + 21i)$

12. $(10 + 15i) - (48 - 30i)$

13. $(28 - 4i) - (10 - 30i)$

14. $(6 - 4i)(6 + 4i)$

15. $(8 - 11i)(8 - 11i)$

16. $(4 + 3i)(2 - 5i)$

17. $(7 + 2i)(9 - 6i)$

18. $\frac{6 + 5i}{-2i}$

19. $\frac{2}{7 - 8i}$

20. $\frac{3 - i}{2 - i}$

21. $\frac{2 - 4i}{1 + 3i}$

Solve each equation.

22. $5n^2 + 35 = 0$

23. $2m^2 + 10 = 0$

24. $4m^2 + 76 = 0$

25. $-2m^2 - 6 = 0$

26. $-5m^2 - 65 = 0$

27. $\frac{3}{4}x^2 + 12 = 0$

Find the values of ℓ and m that make each equation true.

28. $15 - 28i = 3\ell + (4m)i$

29. $(6 - \ell) + (3m)i = -12 + 27i$

30. $(3\ell + 4) + (3 - m)i = 16 - 3i$

31. $(7 + m) + (4\ell - 10)i = 3 - 6i$

32. ELECTRICITY The impedance in one part of a series circuit is $1 + 3j$ ohms and the impedance in another part of the circuit is $7 - 5j$ ohms. Add these complex numbers to find the total impedance in the circuit.

33. ELECTRICITY Using the formula $E = IZ$, find the voltage E in a circuit when the current I is $3 - j$ amps and the impedance Z is $3 + 2j$ ohms.

NAME ______________________ DATE ____________ PERIOD ________

5-5 Skills Practice

Completing the Square

Solve each equation by using the Square Root Property. Round to the nearest hundredth if necessary.

1. $x^2 - 8x + 16 = 1$

2. $x^2 + 4x + 4 = 1$

3. $x^2 + 12x + 36 = 25$

4. $4x^2 - 4x + 1 = 9$

5. $x^2 + 4x + 4 = 2$

6. $x^2 - 2x + 1 = 5$

7. $x^2 - 6x + 9 = 7$

8. $x^2 + 16x + 64 = 15$

Find the value of *c* that makes each trinomial a perfect square. Then write the trinomial as a perfect square.

9. $x^2 + 10x + c$

10. $x^2 - 14x + c$

11. $x^2 + 24x + c$

12. $x^2 + 5x + c$

13. $x^2 - 9x + c$

14. $x^2 - x + c$

Solve each equation by completing the square.

15. $x^2 - 13x + 36 = 0$

16. $x^2 + 3x = 0$

17. $x^2 + x - 6 = 0$

18. $x^2 - 4x - 13 = 0$

19. $2x^2 + 7x - 4 = 0$

20. $3x^2 + 2x - 1 = 0$

21. $x^2 + 3x - 6 = 0$

22. $x^2 - x - 3 = 0$

23. $x^2 = -11$

24. $x^2 - 2x + 4 = 0$

NAME ______________________ DATE ____________ PERIOD ____________

5-5 Practice

Completing the Square

Solve each equation by using the Square Root Property. Round to the nearest hundredth if necessary.

1. $x^2 + 8x + 16 = 1$

2. $x^2 + 6x + 9 = 1$

3. $x^2 + 10x + 25 = 16$

4. $x^2 - 14x + 49 = 9$

5. $4x^2 + 12x + 9 = 4$

6. $x^2 - 8x + 16 = 8$

7. $x^2 - 6x + 9 = 5$

8. $x^2 - 2x + 1 = 2$

9. $9x^2 - 6x + 1 = 2$

Find the value of c that makes each trinomial a perfect square. Then write the trinomial as a perfect square.

10. $x^2 + 12x + c$

11. $x^2 - 20x + c$

12. $x^2 + 11x + c$

13. $x^2 + 0.8x + c$

14. $x^2 - 2.2x + c$

15. $x^2 - 0.36x + c$

16. $x^2 + \frac{5}{6}x + c$

17. $x^2 - \frac{1}{4}x + c$

18. $x^2 - \frac{5}{3}x + c$

Solve each equation by completing the square.

19. $x^2 + 6x + 8 = 0$

20. $3x^2 + x - 2 = 0$

21. $3x^2 - 5x + 2 = 0$

22. $x^2 + 18 = 9x$

23. $x^2 - 14x + 19 = 0$

24. $x^2 + 16x - 7 = 0$

25. $2x^2 + 8x - 3 = 0$

26. $x^2 + x - 5 = 0$

27. $2x^2 - 10x + 5 = 0$

28. $x^2 + 3x + 6 = 0$

29. $2x^2 + 5x + 6 = 0$

30. $7x^2 + 6x + 2 = 0$

31. GEOMETRY When the dimensions of a cube are reduced by 4 inches on each side, the surface area of the new cube is 864 square inches. What were the dimensions of the original cube?

32. INVESTMENTS The amount of money A in an account in which P dollars are invested for 2 years is given by the formula $A = P(1 + r)^2$, where r is the interest rate compounded annually. If an investment of $800 in the account grows to $882 in two years, at what interest rate was it invested?

NAME ______________________ DATE ____________ PERIOD ____________

5-6 Skills Practice

The Quadratic Formula and the Discriminant

Complete parts a–c for each quadratic equation.

a. Find the value of the discriminant.

b. Describe the number and type of roots.

c. Find the exact solutions by using the Quadratic Formula.

1. $x^2 - 8x + 16 = 0$

2. $x^2 - 11x - 26 = 0$

3. $3x^2 - 2x = 0$

4. $20x^2 + 7x - 3 = 0$

5. $5x^2 - 6 = 0$

6. $x^2 - 6 = 0$

7. $x^2 + 8x + 13 = 0$

8. $5x^2 - x - 1 = 0$

9. $x^2 - 2x - 17 = 0$

10. $x^2 + 49 = 0$

11. $x^2 - x + 1 = 0$

12. $2x^2 - 3x = -2$

Solve each equation by using the Quadratic Formula.

13. $x^2 = 64$

14. $x^2 - 30 = 0$

15. $x^2 - x = 30$

16. $16x^2 - 24x - 27 = 0$

17. $x^2 - 4x - 11 = 0$

18. $x^2 - 8x - 17 = 0$

19. $x^2 + 25 = 0$

20. $3x^2 + 36 = 0$

21. $2x^2 + 10x + 11 = 0$

22. $2x^2 - 7x + 4 = 0$

23. $8x^2 + 1 = 4x$

24. $2x^2 + 2x + 3 = 0$

25. PARACHUTING Ignoring wind resistance, the distance $d(t)$ in feet that a parachutist falls in t seconds can be estimated using the formula $d(t) = 16t^2$. If a parachutist jumps from an airplane and falls for 1100 feet before opening her parachute, how many seconds pass before she opens the parachute?

NAME ______________________ DATE __________ PERIOD __________

5-6 Practice

The Quadratic Formula and the Discriminant

Solve each equation by using the Quadratic Formula.

1. $7x^2 - 5x = 0$

2. $4x^2 - 9 = 0$

3. $3x^2 + 8x = 3$

4. $x^2 - 21 = 4x$

5. $3x^2 - 13x + 4 = 0$

6. $15x^2 + 22x = -8$

7. $x^2 - 6x + 3 = 0$

8. $x^2 - 14x + 53 = 0$

9. $3x^2 = -54$

10. $25x^2 - 20x - 6 = 0$

11. $4x^2 - 4x + 17 = 0$

12. $8x - 1 = 4x^2$

13. $x^2 = 4x - 15$

14. $4x^2 - 12x + 7 = 0$

Complete parts a–c for each quadratic equation.

a. Find the value of the discriminant.

b. Describe the number and type of roots.

c. Find the exact solutions by using the Quadratic Formula.

15. $x^2 - 16x + 64 = 0$

16. $x^2 = 3x$

17. $9x^2 - 24x + 16 = 0$

18. $x^2 - 3x = 40$

19. $3x^2 + 9x - 2 = 0$

20. $2x^2 + 7x = 0$

21. $5x^2 - 2x + 4 = 0$

22. $12x^2 - x - 6 = 0$

23. $7x^2 + 6x + 2 = 0$

24. $12x^2 + 2x - 4 = 0$

25. $6x^2 - 2x - 1 = 0$

26. $x^2 + 3x + 6 = 0$

27. $4x^2 - 3x^2 - 6 = 0$

28. $16x^2 - 8x + 1 = 0$

29. $2x^2 - 5x - 6 = 0$

30. GRAVITATION The height $h(t)$ in feet of an object t seconds after it is propelled straight up from the ground with an initial velocity of 60 feet per second is modeled by the equation $h(t) = -16t^2 + 60t$. At what times will the object be at a height of 56 feet?

31. STOPPING DISTANCE The formula $d = 0.05s^2 + 1.1s$ estimates the minimum stopping distance d in feet for a car traveling s miles per hour. If a car stops in 200 feet, what is the fastest it could have been traveling when the driver applied the brakes?

NAME ______________________ DATE __________ PERIOD __________

5-7 Skills Practice

Transformations with Quadratic Functions

Write each quadratic function in vertex form. Then identify the vertex, axis of symmetry, and direction of opening.

1. $y = (x - 2)^2$

2. $y = -x^2 + 4$

3. $y = x^2 - 6$

4. $y = -3(x + 5)^2$

5. $y = -5x^2 + 9$

6. $y = (x - 2)^2 - 18$

7. $y = x^2 - 2x - 5$

8. $y = x^2 + 6x + 2$

9. $y = -3x^2 + 24x$

Graph each function.

10. $y = (x - 3)^2 - 1$

11. $y = (x + 1)^2 + 2$

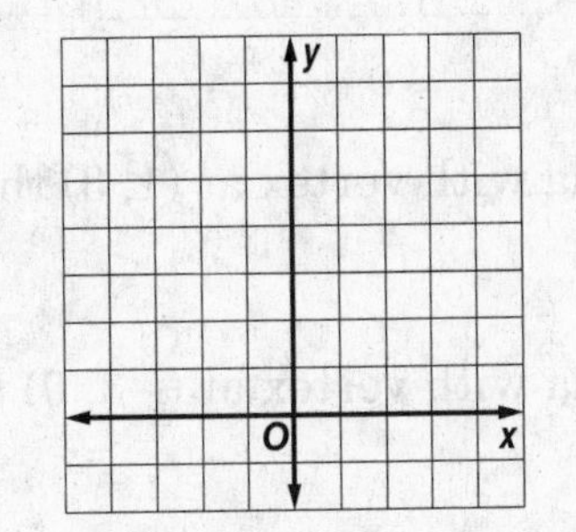

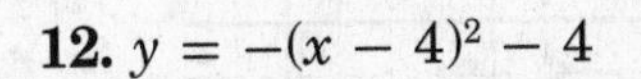

12. $y = -(x - 4)^2 - 4$

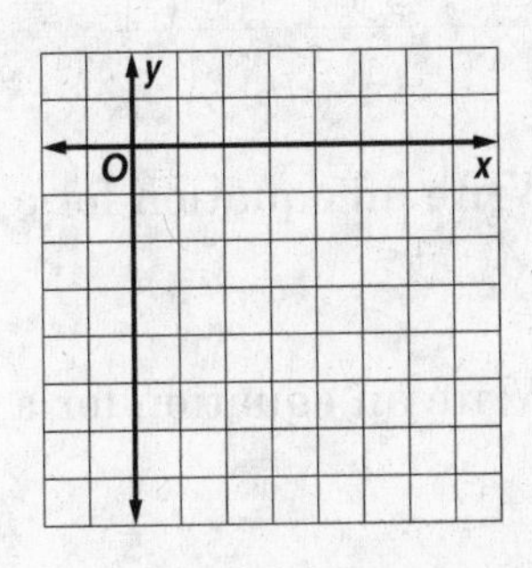

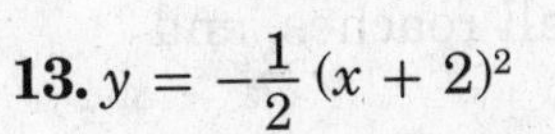

13. $y = -\frac{1}{2}(x + 2)^2$

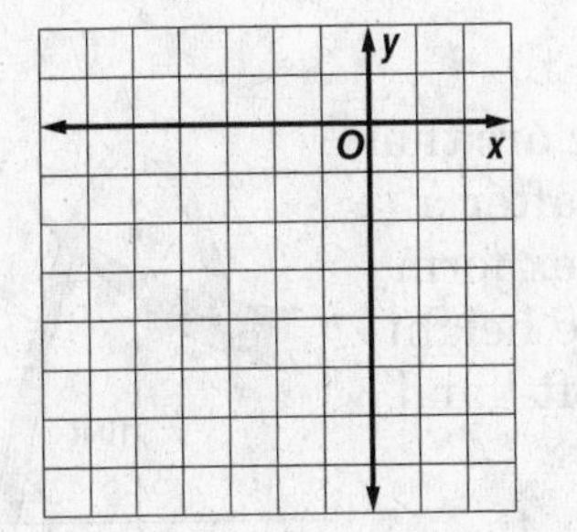

14. $y = -3x^2 + 4$

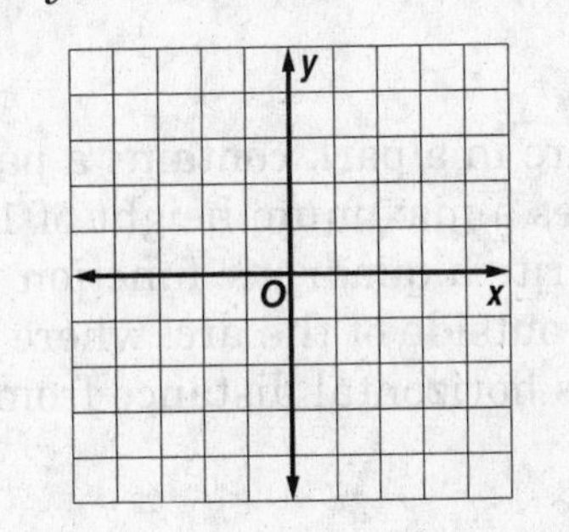

15. $y = x^2 + 6x + 4$

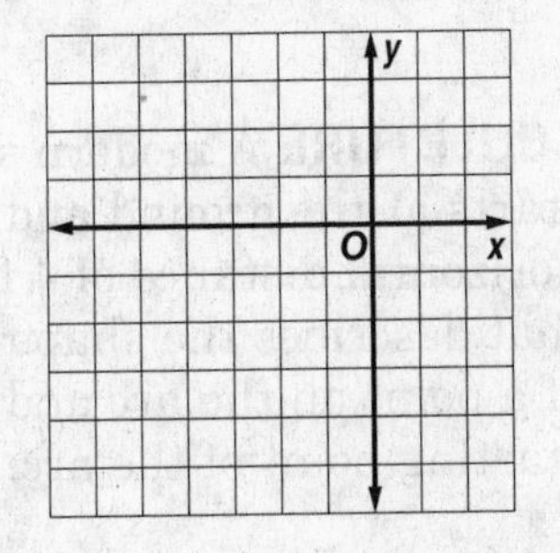

5-7 Practice

Transformations with Quadratic Functions

Write each equation in vertex form. Then identify the vertex, axis of symmetry, and direction of opening.

1. $y = -6x^2 - 24x - 25$
2. $y = 2x^2 + 2$
3. $y = -4x^2 + 8x$
4. $y = x^2 + 10x + 20$
5. $y = 2x^2 + 12x + 18$
6. $y = 3x^2 - 6x + 5$
7. $y = -2x^2 - 16x - 32$
8. $y = -3x^2 + 18x - 21$
9. $y = 2x^2 + 16x + 29$

Graph each function.

10. $y = (x + 3)^2 - 1$ 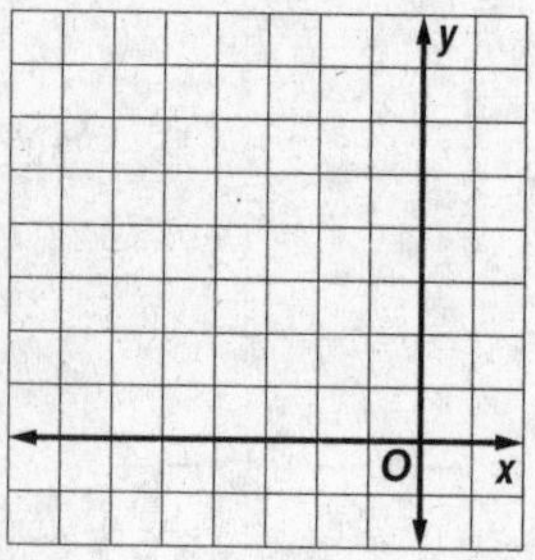
11. $y = -x^2 + 6x - 5$ 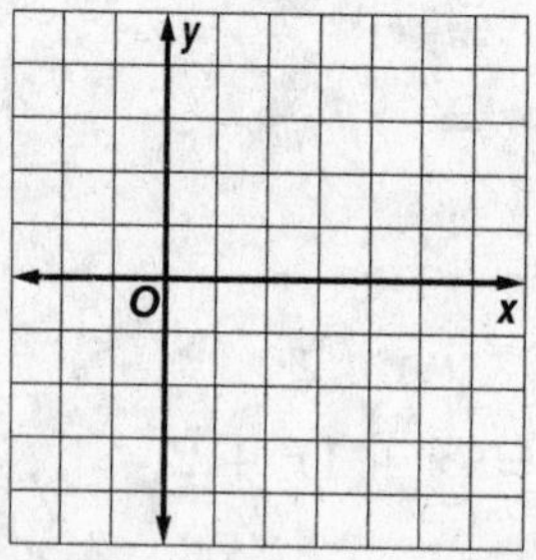
12. $y = 2x^2 - 2x + 1$

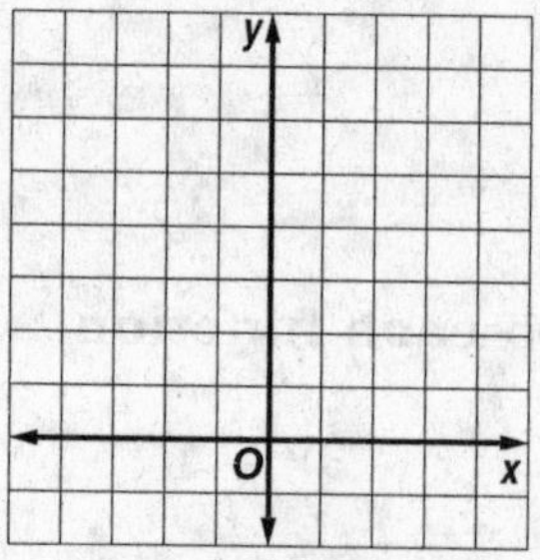

13. Write an equation for a parabola with vertex at (1, 3) that passes through (−2, −15).

14. Write an equation for a parabola with vertex at (−3, 0) that passes through (3, 18).

15. **BASEBALL** The height h of a baseball t seconds after being hit is given by $h(t) = -16t^2 + 80t + 3$. What is the maximum height that the baseball reaches, and when does this occur?

16. **SCULPTURE** A modern sculpture in a park contains a parabolic arc that starts at the ground and reaches a maximum height of 10 feet after a horizontal distance of 4 feet. Write a quadratic function in vertex form that describes the shape of the outside of the arc, where y is the height of a point on the arc and x is its horizontal distance from the left-hand starting point of the arc.

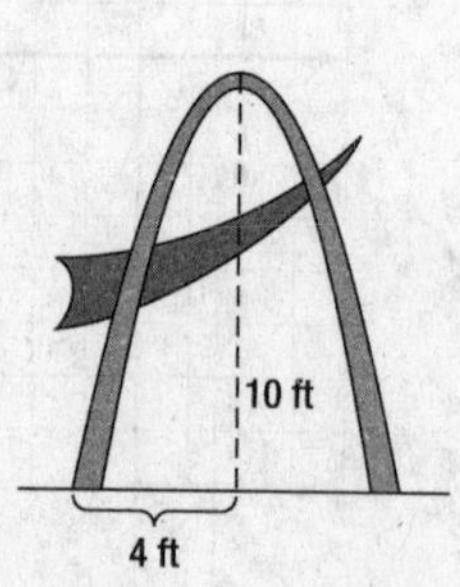

NAME ______________________________ DATE ____________ PERIOD ____________

5-8 Skills Practice

Quadratic Inequalities

Graph each inequality.

1. $y \geq x^2 - 4x + 4$

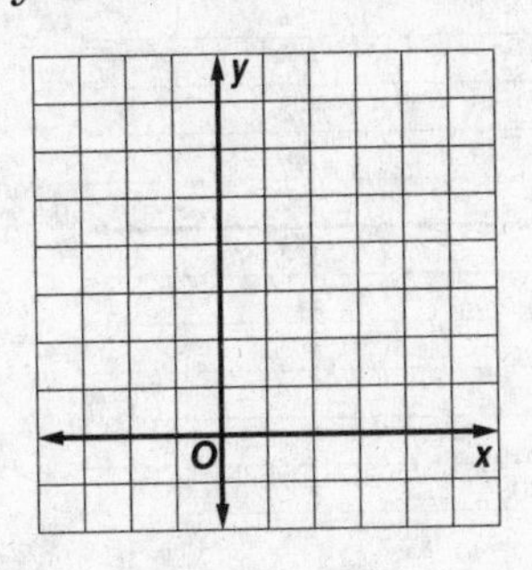

2. $y \leq x^2 - 4$

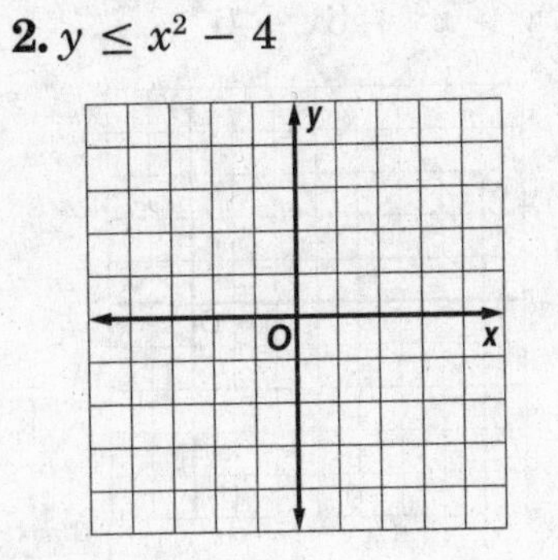

3. $y > x^2 + 2x - 5$

Solve each inequality by graphing.

4. $x^2 - 6x + 9 \leq 0$

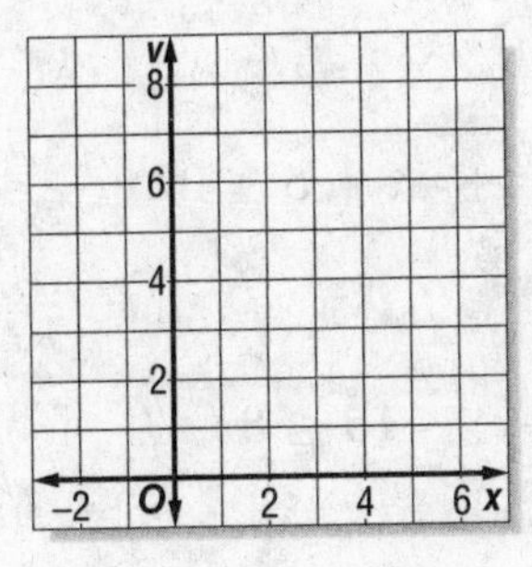

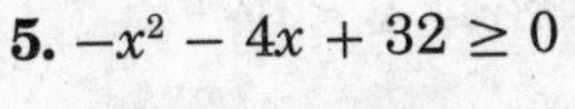

5. $-x^2 - 4x + 32 \geq 0$

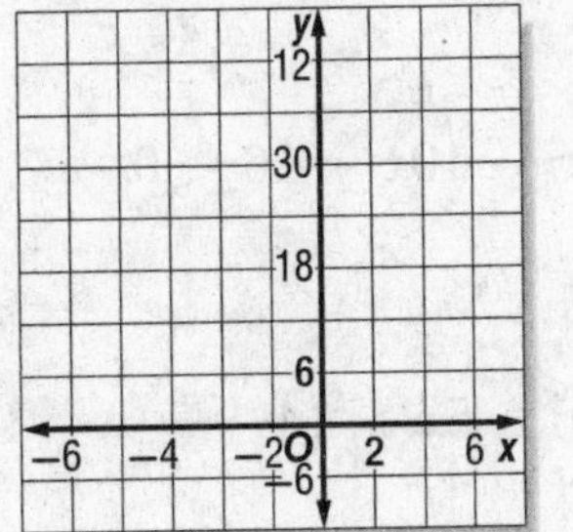

6. $x^2 + x - 10 > 10$

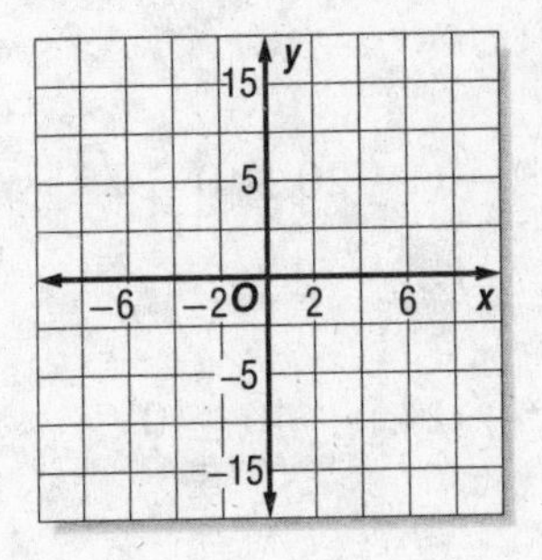

Solve each inequality algebraically.

7. $x^2 - 3x - 10 < 0$

8. $x^2 + 2x - 35 \geq 0$

9. $x^2 - 18x + 81 \leq 0$

10. $x^2 \leq 36$

11. $x^2 - 7x > 0$

12. $x^2 + 7x + 6 < 0$

13. $x^2 + x - 12 > 0$

14. $x^2 + 9x + 18 \leq 0$

15. $x^2 - 10x + 25 \geq 0$

16. $-x^2 - 2x + 15 \geq 0$

17. $x^2 + 3x > 0$

18. $2x^2 + 2x > 4$

19. $-x^2 - 64 \leq -16x$

20. $9x^2 + 12x + 9 < 0$

NAME ______________________ DATE ____________ PERIOD ____________

5-8 Practice

Quadratic Inequalities

Graph each inequality.

1. $y \leq x^2 + 4$

2. $y > x^2 + 6x + 6$

3. $y < 2x^2 - 4x - 2$

Solve each inequality.

4. $x^2 + 2x + 1 > 0$

5. $x^2 - 3x + 2 \leq 0$

6. $x^2 + 10x + 7 \geq 0$

7. $x^2 - x - 20 > 0$

8. $x^2 - 10x + 16 < 0$

9. $x^2 + 4x + 5 \leq 0$

10. $x^2 + 14x + 49 \geq 0$

11. $x^2 - 5x > 14$

12. $-x^2 - 15 \leq 8x$

13. $-x^2 + 5x - 7 \leq 0$

14. $9x^2 + 36x + 36 \leq 0$

15. $9x \leq 12x^2$

16. $4x^2 + 4x + 1 > 0$

17. $5x^2 + 10 \geq 27x$

18. $9x^2 + 31x + 12 \leq 0$

19. FENCING Vanessa has 180 feet of fencing that she intends to use to build a rectangular play area for her dog. She wants the play area to enclose at least 1800 square feet. What are the possible widths of the play area?

20. BUSINESS A bicycle maker sold 300 bicycles last year at a profit of $300 each. The maker wants to increase the profit margin this year, but predicts that each $20 increase in profit will reduce the number of bicycles sold by 10. How many $20 increases in profit can the maker add in and expect to make a total profit of at least $100,000?

NAME ______________________ DATE ____________ PERIOD ____________

6-1 Skills Practice

Operations with Polynomials

Simplify. Assume that no variable equals 0.

1. $b^4 \cdot b^3$

2. $c^5 \cdot c^2 \cdot c^2$

3. $a^{-4} \cdot a^{-3}$

4. $x^5 \cdot x^{-4} \cdot x$

5. $(2x)^2(4y)^2$

6. $-2gh(g^3h^5)$

7. $10x^2y^3(10xy^8)$

8. $\frac{24wz^7}{3w^3z^5}$

9. $\frac{-6a^4bc^8}{36a^7b^2c}$

10. $\frac{-10pt^4r}{-5p^3t^2r}$

11. $(g + 5) + (2g + 7)$

12. $(5d + 5) - (d + 1)$

13. $(x^2 - 3x - 3) + (2x^2 + 7x - 2)$

14. $(-2f^2 - 3f - 5) + (-2f^2 - 3f + 8)$

15. $-5(2c^2 - d^2)$

16. $x^2(2x + 9)$

17. $(a - 5)^2$

18. $(2x - 3)(3x - 5)$

19. $(r - 2t)(r + 2t)$

20. $(3y + 4)(2y - 3)$

21. $(3 - 2b)(3 + 2b)$

22. $(3w + 1)^2$

NAME ______________________ DATE ____________ PERIOD ____________

6-1 Practice

Operations with Polynomials

Simplify. Assume that no variable equals 0.

1. $n^5 \cdot n^2$

2. $y^7 \cdot y^3 \cdot y^2$

3. $t^9 \cdot t^{-8}$

4. $x^{-4} \cdot x^{-4} \cdot x^4$

5. $(2f^4)^6$

6. $(-2b^{-2}c^3)^3$

7. $(4d^2t^5v^{-4})(-5dt^{-3}v^{-1})$

8. $8u(2z)^3$

9. $\frac{12m^8y^6}{-9my^4}$

10. $\frac{-6n^5x^3}{18nx^7}$

11. $\frac{-27x^3(-x^7)}{16x^4}$

12. $\left(\frac{2}{3r^2t^3z^6}\right)^2$

13. $-(4w^{-3}z^{-5})(8w)^2$

14. $(m^4n^6)^4(m^3n^2p^5)^6$

15. $\left(\frac{3}{2}d^{-2}f^4\right)^4\left(-\frac{4}{3}d^5f\right)^3$

16. $\left(\frac{2x^3y^2}{-x^2y^5}\right)^{-2}$

17. $\frac{(3x^{-2}y^3)(5xy^{-8})}{(x^{-3})^4y^{-2}}$

18. $\frac{-20(m^2v)(-v)^3}{5(-v)^2(-m^4)}$

19. $(3n^2 + 1) + (8n^2 - 8)$

20. $(6w - 11w^2) - (4 + 7w^2)$

21. $(w + 2t)(w^2 - 2wt + 4t^2)$

22. $(x + y)(x^2 - 3xy + 2y^2)$

23. BANKING Terry invests $1500 in two mutual funds. The first year, one fund grows 3.8% and the other grows 6%. Write a polynomial to represent the amount Terry's $1500 grows to in that year if x represents the amount he invested in the fund with the lesser growth rate.

24. GEOMETRY The area of the base of a rectangular box measures $2x^2 + 4x - 3$ square units. The height of the box measures x units. Find a polynomial expression for the volume of the box.

NAME ______________________ DATE ____________ PERIOD ____________

6-2 Skills Practice

Dividing Polynomials

Simplify.

1. $\frac{10c + 6}{2}$

2. $\frac{12x + 20}{4}$

3. $\frac{15y^3 + 6y^2 + 3y}{3y}$

4. $\frac{12x^2 - 4x - 8}{4x}$

5. $(15q^6 + 5q^2)(5q^4)^{-1}$

6. $(4f^5 - 6f^4 + 12f^3 - 8f^2)(4f^2)^{-1}$

7. $(6j^2k - 9jk^2) \div 3jk$

8. $(4a^2h^2 - 8a^3h + 3a^4) \div (2a^2)$

9. $(n^2 + 7n + 10) \div (n + 5)$

10. $(d^2 + 4d + 3) \div (d + 1)$

11. $(2t^2 + 13t + 15) \div (t + 5)$

12. $(6y^2 + y - 2)(2y - 1)^{-1}$

13. $(4g^2 - 9) \div (2g + 3)$

14. $(2x^2 - 5x - 4) \div (x - 3)$

15. $\frac{u^2 + 5u - 12}{u - 3}$

16. $\frac{6x^3 + 5x^2 + 9}{2x + 3}$

17. $(3v^2 - 7v - 10)(v - 4)^{-1}$

18. $(3t^4 + 4t^3 - 32t^2 - 5t - 20)(t + 4)^{-1}$

19. $\frac{y^3 - y^2 - 6}{y + 2}$

20. $\frac{2x^3 - x^2 - 19x + 15}{x - 3}$

21. $(4p^3 - 3p^2 + 2p) \div (p - 1)$

22. $(3c^4 + 6c^3 - 2c + 4)(c + 2)^{-1}$

23. GEOMETRY The area of a rectangle is $x^3 + 8x^2 + 13x - 12$ square units. The width of the rectangle is $x + 4$ units. What is the length of the rectangle?

NAME ______________________ DATE ____________ PERIOD ________

6-2 Practice

Dividing Polynomials

Simplify.

1. $\frac{15r^{10} - 5r^8 + 40r^2}{5r^4}$

2. $\frac{6k^2m - 12k^3m^2 + 9m^3}{2km^2}$

3. $(-30x^3y + 12x^2y^2 - 18x^2y) \div (-6x^2y)$

4. $(-6w^3z^4 - 3w^2z^5 + 4w + 5z) \div (2w^2z)$

5. $(4a^3 - 8a^2 + a^2)(4a)^{-1}$

6. $(28d^3k^2 + d^2k^2 - 4dk^2)(4dk^2)^{-1}$

7. $\frac{f^2 + 7f + 10}{f + 2}$

8. $\frac{2x^3 + 3x - 14}{x - 2}$

9. $(a^3 - 64) \div (a - 4)$

10. $(b^3 + 27) \div (b + 3)$

11. $\frac{2x^3 + 6x + 152}{x + 4}$

12. $\frac{2x^3 + 4x - 6}{x + 3}$

13. $(3w^3 + 7w^2 - 4w + 3) \div (w + 3)$

14. $(6y^4 + 15y^3 - 28y - 6) \div (y + 2)$

15. $(x^4 - 3x^3 - 11x^2 + 3x + 10) \div (x - 5)$

16. $(3m^5 + m - 1) \div (m + 1)$

17. $(x^4 - 3x^3 + 5x - 6)(x + 2)^{-1}$

18. $(6y^2 - 5y - 15)(2y + 3)^{-1}$

19. $\frac{4x^2 - 2x + 6}{2x - 3}$

20. $\frac{6x^2 - x - 7}{3x + 1}$

21. $(2r^3 + 5r^2 - 2r - 15) \div (2r - 3)$

22. $(6t^3 + 5t^2 - 2t + 1) \div (3t + 1)$

23. $\frac{4p^4 - 17p^2 + 14p - 3}{2p - 3}$

24. $\frac{2h^4 - h^3 + h^2 + h - 3}{h^2 - 1}$

25. GEOMETRY The area of a rectangle is $2x^2 - 11x + 15$ square feet. The length of the rectangle is $2x - 5$ feet. What is the width of the rectangle?

26. GEOMETRY The area of a triangle is $15x^4 + 3x^3 + 4x^2 - x - 3$ square meters. The length of the base of the triangle is $6x^2 - 2$ meters. What is the height of the triangle?

NAME ______________________ DATE __________ PERIOD ______

6-3 Skills Practice

Polynomial Functions

State the degree and leading coefficient of each polynomial in one variable. If it is not a polynomial in one variable, explain why.

1. $a + 8$

2. $(2x - 1)(4x^2 + 3)$

3. $-5x^5 + 3x^3 - 8$

4. $18 - 3y + 5y^2 - y^5 + 7y^6$

5. $u^3 + 4u^2t^2 + t^4$

6. $2r - r^2 + \frac{1}{r^2}$

Find $p(-1)$ and $p(2)$ for each function.

7. $p(x) = 4 - 3x$

8. $p(x) = 3x + x^2$

9. $p(x) = 2x^2 - 4x + 1$

10. $p(x) = -2x^3 + 5x + 3$

11. $p(x) = x^4 + 8x^2 - 10$

12. $p(x) = \frac{1}{3}x^2 - \frac{2}{3}x + 2$

If $p(x) = 4x^2 - 3$ and $r(x) = 1 + 3x$, find each value.

13. $p(a)$

14. $r(2a)$

15. $3r(a)$

16. $-4p(a)$

17. $p(a^2)$

18. $r(x + 2)$

For each graph,
a. describe the end behavior,
b. determine whether it represents an odd-degree or an even-degree function, and
c. state the number of real zeroes.

19.

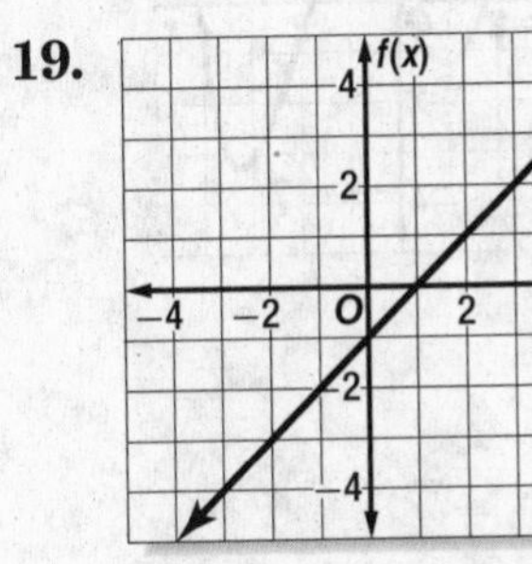

20.

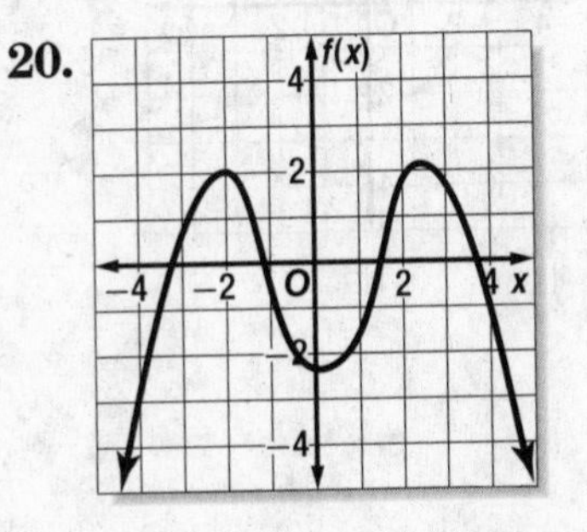

21.

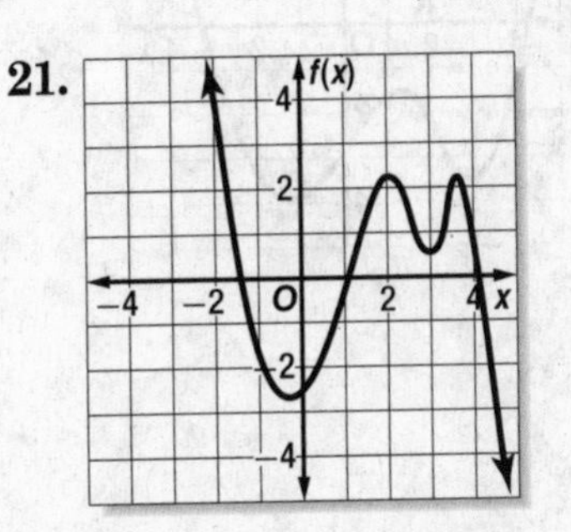

NAME ______________________________ DATE ____________ PERIOD ________

6-3 Practice

Polynomial Functions

State the degree and leading coefficient of each polynomial in one variable. If it is not a polynomial in one variable, explain why.

1. $(3x^2 + 1)(2x^2 - 9)$

2. $\frac{1}{5}a^3 - \frac{3}{5}a^2 + \frac{4}{5}a$

3. $\frac{2}{m^2} + 3m - 12$

4. $27 + 3xy^3 - 12x^2y^2 - 10y$

Find $p(-2)$ and $p(3)$ for each function.

5. $p(x) = x^3 - x^5$

6. $p(x) = -7x^2 + 5x + 9$

7. $p(x) = -x^5 + 4x^3$

8. $p(x) = 3x^3 - x^2 + 2x - 5$

9. $p(x) = x^4 + \frac{1}{2}x^3 - \frac{1}{2}x$

10. $p(x) = \frac{1}{3x^3} + \frac{2}{3x^2} + 3x$

If $p(x) = 3x^2 - 4$ and $r(x) = 2x^2 - 5x + 1$, find each value.

11. $p(8a)$

12. $r(a^2)$

13. $-5r(2a)$

14. $r(x + 2)$

15. $p(x^2 - 1)$

16. $5p(x + 2)]$

For each graph,
a. describe the end behavior,
b. determine whether it represents an odd-degree or an even-degree function, and
c. state the number of real zeroes.

17. 18. 19.

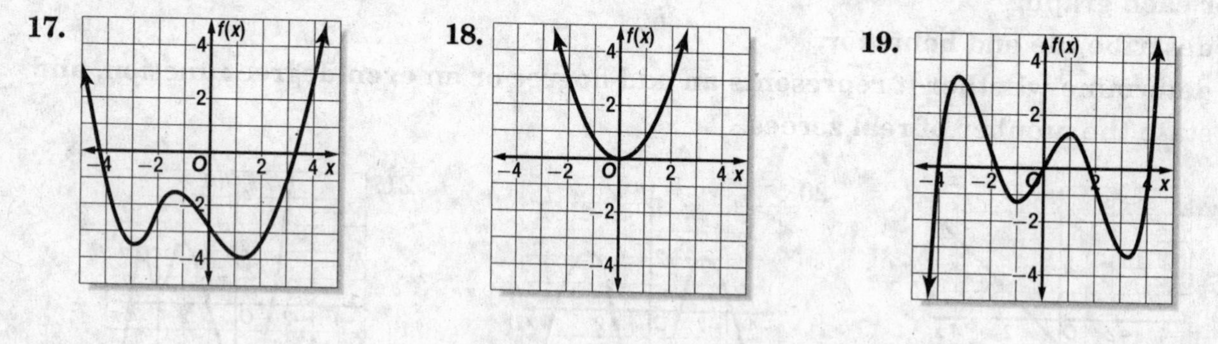

20. **WIND CHILL** The function $C(w) = 0.013w^2 - w - 7$ estimates the wind chill temperature $C(w)$ at 0°F for wind speeds w from 5 to 30 miles per hour. Estimate the wind chill temperature at 0°F if the wind speed is 20 miles per hour.

NAME ________________________ DATE ____________ PERIOD ________

6-4 Skills Practice

Analyzing Graphs of Polynomial Functions

Complete each of the following.

a. Graph each function by making a table of values.

b. Determine the consecutive values of x between which each real zero is located.

c. Estimate the x-coordinates at which the relative maxima and minima occur.

1. $f(x) = x^3 - 3x^2 + 1$

x	f(x)
−2	
−1	
0	
1	
2	
3	
4	

2. $f(x) = x^3 - 3x + 1$

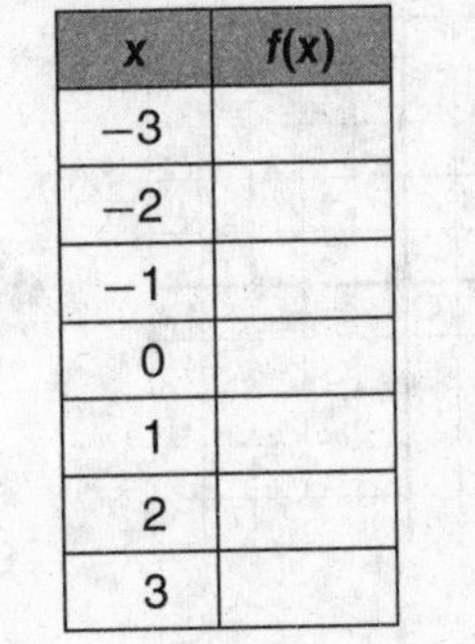

x	f(x)
−3	
−2	
−1	
0	
1	
2	
3	

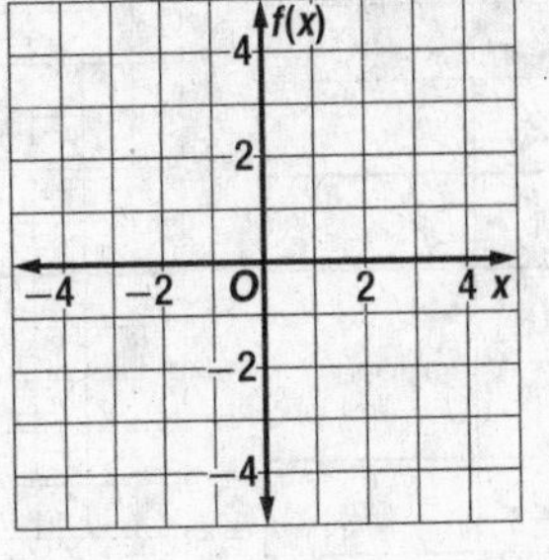

3. $f(x) = 2x^3 + 9x^2 + 12x + 2$

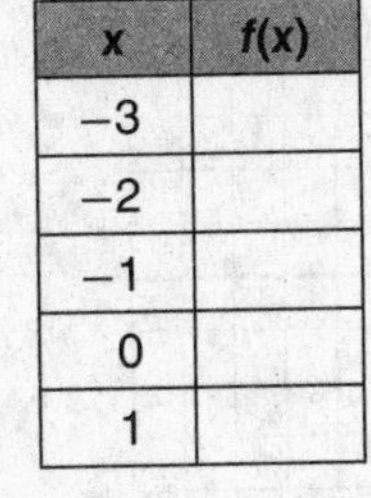

x	f(x)
−3	
−2	
−1	
0	
1	

4. $f(x) = 2x^3 - 3x^2 + 2$

x	f(x)
−1	
0	
1	
2	
3	

5. $f(x) = x^4 - 2x^2 - 2$

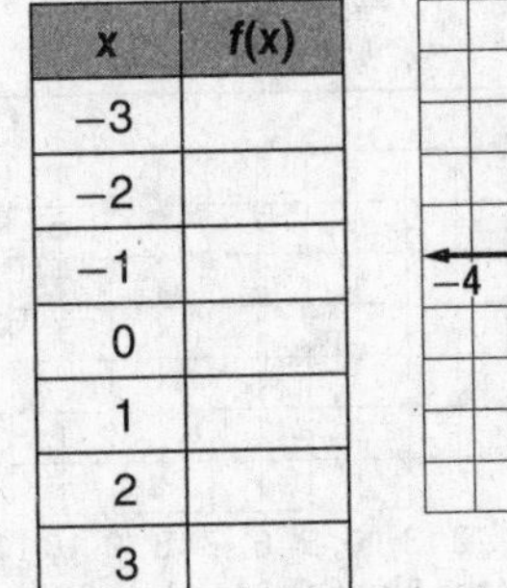

x	f(x)
−3	
−2	
−1	
0	
1	
2	
3	

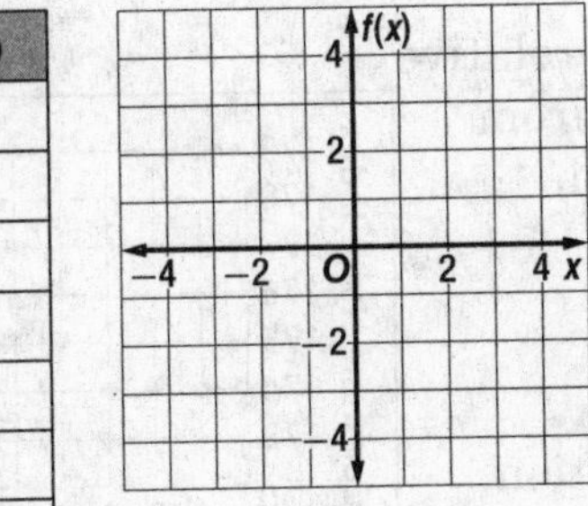

6. $f(x) = 0.5x^4 - 4x^2 + 4$

x	f(x)
−3	
−2	
−1	
0	
1	
2	
3	

NAME _____________________________ DATE ____________ PERIOD ____________

6-4 Practice

Analyzing Graphs of Polynomial Functions

Complete each of the following.

a. Graph each function by making a table of values.

b. Determine the consecutive values of *x* between which each real zero is located.

c. Estimate the *x*-coordinates at which the relative maxima and minima occur.

1. $f(x) = -x^3 + 3x^2 - 3$

x	f(x)
−2	
−1	
0	
1	
2	
3	
4	

2. $f(x) = x^3 - 1.5x^2 - 6x + 1$

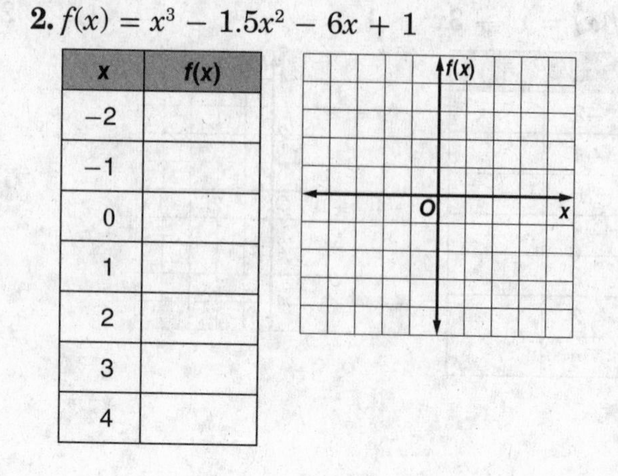

x	f(x)
−2	
−1	
0	
1	
2	
3	
4	

3. $f(x) = 0.75x^4 + x^3 - 3x^2 + 4$

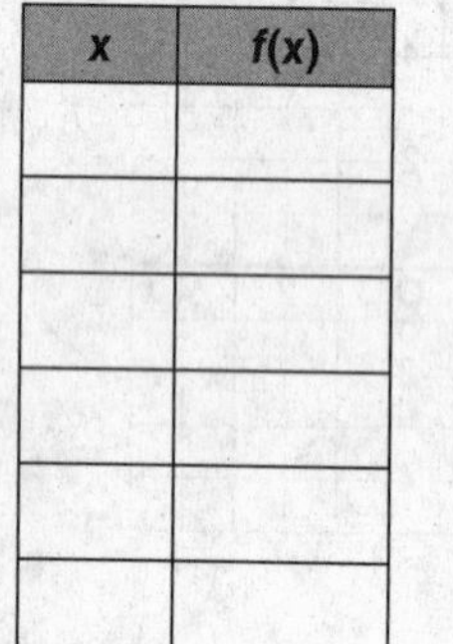

x	f(x)

4. $f(x) = x^4 + 4x^3 + 6x^2 + 4x - 3$

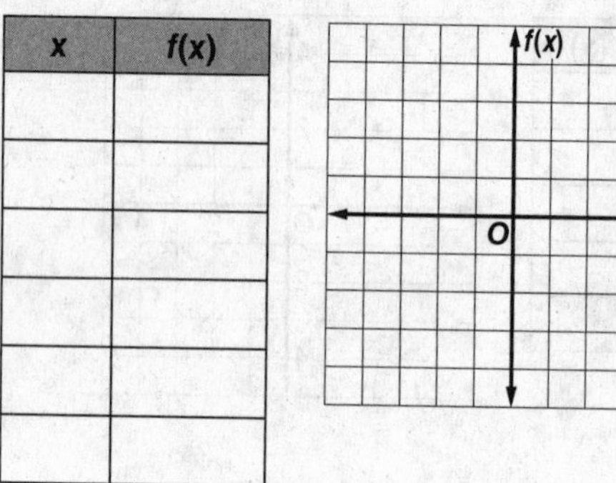

x	f(x)

5. PRICES The Consumer Price Index (CPI) gives the relative price for a fixed set of goods and services. The CPI from September, 2000 to July, 2001 is shown in the graph.

Source: *U. S. Bureau of Labor Statistics*

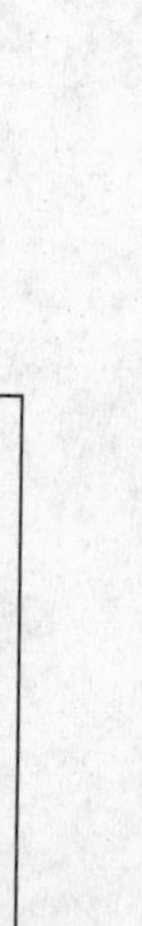

a. Describe the turning points of the graph.

b. If the graph were modeled by a polynomial equation, what is the least degree the equation could have?

6. LABOR A town's jobless rate can be modeled by (1, 3.3), (2, 4.9), (3, 5.3), (4, 6.4), (5, 4.5), (6, 5.6), (7, 2.5), and (8, 2.7). How many turning points would the graph of a polynomial function through these points have? Describe them.

NAME ________________ DATE ________ PERIOD ________

6-5 Skills Practice

Solving Polynomial Equations

Factor completely. If the polynomial is not factorable, write *prime*.

1. $7x^2 - 14x$

2. $19x^3 - 38x^2$

3. $21x^3 - 18x^2y + 24xy^2$

4. $8j^3k - 4jk^3 - 7$

5. $a^2 + 7a - 18$

6. $2ak - 6a + k - 3$

7. $b^2 + 8b + 7$

8. $z^2 - 8z - 10$

9. $4f^2 - 64$

10. $d^2 - 12d + 36$

11. $9x^2 + 25$

12. $y^2 + 18y + 81$

13. $n^3 - 125$

14. $m^4 - 1$

Write each expression in quadratic form, if possible.

15. $5x^4 + 2x^2 - 8$

16. $3y^8 - 4y^2 + 3$

17 $100a^6 + a^3$

18. $x^8 + 4x^4 + 9$

19. $12x^4 - 7x^2$

20. $6b^5 + 3b^3 - 1$

Solve each equation.

21. $a^3 - 9a^2 + 14a = 0$

22. $x^3 = 3x^2$

23. $t^4 - 3t^3 - 40t^2 = 0$

24. $b^3 - 8b^2 + 16b = 0$

NAME ______________________ DATE ____________ PERIOD ____________

6-5 Practice

Solving Polynomial Equations

Factor completely. If the polynomial is not factorable, write *prime*.

1. $15a^2b - 10ab^2$

2. $3st^2 - 9s^3t + 6s^2t^2$

3. $3x^3y^2 - 2x^2y + 5xy$

4. $2x^3y - x^2y + 5xy^2 + xy^3$

5. $21 - 7t + 3r - rt$

6. $x^2 - xy + 2x - 2y$

7. $y^2 + 20y + 96$

8. $4ab + 2a + 6b + 3$

9. $6n^2 - 11n - 2$

10. $6x^2 + 7x - 3$

11. $x^2 - 8x - 8$

12. $6p^2 - 17p - 45$

Write each expression in quadratic form, if possible.

13. $10b^4 + 3b^2 - 11$

14. $-5x^8 + x^2 + 6$

15. $28d^6 + 25d^3$

16. $4c^8 + 4c^4 + 7$

17. $500x^4 - x^2$

18. $8b^5 - 8b^3 - 1$

Solve each equation.

19. $y^4 - 7y^3 - 18y^2 = 0$

20. $k^5 + 4k^4 - 32k^3 = 0$

21. $m^4 - 625 = 0$

22. $n^4 - 49n^2 = 0$

23. $x^4 - 50x^2 + 49 = 0$

24. $t^4 - 21t^2 + 80 = 0$

25. PHYSICS A proton in a magnetic field follows a path on a coordinate grid modeled by the function $f(x) = x^4 - 2x^2 - 15$. What are the x-coordinates of the points on the grid where the proton crosses the x-axis?

26. SURVEYING Vista county is setting aside a large parcel of land to preserve it as open space. The county has hired Meghan's surveying firm to survey the parcel, which is in the shape of a right triangle. The longer leg of the triangle measures 5 miles less than the square of the shorter leg, and the hypotenuse of the triangle measures 13 miles less than twice the square of the shorter leg. The length of each boundary is a whole number. Find the length of each boundary.

6-6 Skills Practice

The Remainder and Factor Theorems

Use synthetic substitution to find $f(2)$ and $f(-1)$ for each function.

1. $f(x) = x^2 + 6x + 5$

2. $f(x) = x^2 - x + 1$

3. $f(x) = x^2 - 2x - 2$

4. $f(x) = x^3 + 2x^2 + 5$

5. $f(x) = x^3 - x^2 - 2x + 3$

6. $f(x) = x^3 + 6x^2 + x - 4$

7. $f(x) = x^3 - 3x^2 + x - 2$

8. $f(x) = x^3 - 5x^2 - x + 6$

9. $f(x) = x^4 + 2x^2 - 9$

10. $f(x) = x^4 - 3x^3 + 2x^2 - 2x + 6$

11. $f(x) = x^5 - 7x^3 - 4x + 10$

12. $f(x) = x^6 - 2x^5 + x^4 + x^3 - 9x^2 - 20$

Given a polynomial and one of its factors, find the remaining factors of the polynomial.

13. $x^3 + 2x^2 - x - 2$; $x + 1$

14. $x^3 + x^2 - 5x + 3$; $x - 1$

15. $x^3 + 3x^2 - 4x - 12$; $x + 3$

16. $x^3 - 6x^2 + 11x - 6$; $x - 3$

17. $x^3 + 2x^2 - 33x - 90$; $x + 5$

18. $x^3 - 6x^2 + 32$; $x - 4$

19. $x^3 - x^2 - 10x - 8$; $x + 2$

20. $x^3 - 19x + 30$; $x - 2$

21. $2x^3 + x^2 - 2x - 1$; $x + 1$

22. $2x^3 + x^2 - 5x + 2$; $x + 2$

23. $3x^3 + 4x^2 - 5x - 2$; $3x + 1$

24. $3x^3 + x^2 + x - 2$; $3x - 2$

NAME ______________________________ DATE ____________ PERIOD ____________

6-6 Practice

The Remainder and Factor Theorems

Use synthetic substitution to find $f(-3)$ and $f(4)$ for each function.

1. $f(x) = x^2 + 2x + 3$

2. $f(x) = x^2 - 5x + 10$

3. $f(x) = x^2 - 5x - 4$

4. $f(x) = x^3 - x^2 - 2x + 3$

5. $f(x) = x^3 + 2x^2 + 5$

6. $f(x) = x^3 - 6x^2 + 2x$

7. $f(x) = x^3 - 2x^2 - 2x + 8$

8. $f(x) = x^3 - x^2 + 4x - 4$

9. $f(x) = x^3 + 3x^2 + 2x - 50$

10. $f(x) = x^4 + x^3 - 3x^2 - x + 12$

11. $f(x) = x^4 - 2x^2 - x + 7$

12. $f(x) = 2x^4 - 3x^3 + 4x^2 - 2x + 1$

13. $f(x) = 2x^4 - x^3 + 2x^2 - 26$

14. $f(x) = 3x^4 - 4x^3 + 3x^2 - 5x - 3$

15. $f(x) = x^5 + 7x^3 - 4x - 10$

16. $f(x) = x^6 + 2x^5 - x^4 + x^3 - 9x^2 + 20$

Given a polynomial and one of its factors, find the remaining factors of the polynomial.

17. $x^3 + 3x^2 - 6x - 8;\ x - 2$

18. $x^3 + 7x^2 + 7x - 15;\ x - 1$

19. $x^3 - 9x^2 + 27x - 27;\ x - 3$

20. $x^3 - x^2 - 8x + 12;\ x + 3$

21. $x^3 + 5x^2 - 2x - 24;\ x - 2$

22. $x^3 - x^2 - 14x + 24;\ x + 4$

23. $3x^3 - 4x^2 - 17x + 6;\ x + 2$

24. $4x^3 - 12x^2 - x + 3;\ x - 3$

25. $18x^3 + 9x^2 - 2x - 1;\ 2x + 1$

26. $6x^3 + 5x^2 - 3x - 2;\ 3x - 2$

27. $x^5 + x^4 - 5x^3 - 5x^2 + 4x + 4;\ x + 1$

28. $x^5 - 2x^4 + 4x^3 - 8x^2 - 5x + 10;\ x - 2$

29. POPULATION The projected population in thousands for a city over the next several years can be estimated by the function $P(x) = x^3 + 2x^2 - 8x + 520$, where x is the number of years since 2005. Use synthetic substitution to estimate the population for 2010.

30. VOLUME The volume of water in a rectangular swimming pool can be modeled by the polynomial $2x^3 - 9x^2 + 7x + 6$. If the depth of the pool is given by the polynomial $2x + 1$, what polynomials express the length and width of the pool?

NAME ______________________ DATE ____________ PERIOD ________

6-7 Skills Practice

Roots and Zeros

Solve each equation. State the number and type of roots.

1. $5x + 12 = 0$

2. $x^2 - 4x + 40 = 0$

3. $x^5 + 4x^3 = 0$

4. $x^4 - 625 = 0$

5. $4x^2 - 4x - 1 = 0$

6. $x^5 - 81x = 0$

State the possible number of positive real zeros, negative real zeros, and imaginary zeros of each function.

7. $g(x) = 3x^3 - 4x^2 - 17x + 6$

8. $h(x) = 4x^3 - 12x^2 - x + 3$

9. $f(x) = x^3 - 8x^2 + 2x - 4$

10. $p(x) = x^3 - x^2 + 4x - 6$

11. $q(x) = x^4 + 7x^2 + 3x - 9$

12. $f(x) = x^4 - x^3 - 5x^2 + 6x + 1$

Find all the zeros of each function.

13. $h(x) = x^3 - 5x^2 + 5x + 3$

14. $g(x) = x^3 - 6x^2 + 13x - 10$

15. $h(x) = x^3 + 4x^2 + x - 6$

16. $q(x) = x^3 + 3x^2 - 6x - 8$

17. $g(x) = x^4 - 3x^3 - 5x^2 + 3x + 4$

18. $f(x) = x^4 - 21x^2 + 80$

Write a polynomial function of least degree with integral coefficients that have the given zeros.

19. $-3, -5, 1$

20. $3i$

21. $-5 + i$

22. $-1, \sqrt{3}, -\sqrt{3}$

23. $i, 5i$

24. $-1, 1, i\sqrt{6}$

NAME ______________________________ DATE ______________ PERIOD ______________

6-7 Practice

Roots and Zeros

Solve each equation. State the number and type of roots.

1. $-9x - 15 = 0$

2. $x^4 - 5x^2 + 4 = 0$

3. $x^5 - 81x = 0$

4. $x^3 + x^2 - 3x - 3 = 0$

5. $x^3 + 6x + 20 = 0$

6. $x^4 - x^3 - x^2 - x - 2 = 0$

State the possible number of positive real zeros, negative real zeros, and imaginary zeros of each function.

7. $f(x) = 4x^3 - 2x^2 + x + 3$

8. $p(x) = 2x^4 - 2x^3 + 2x^2 - x - 1$

9. $q(x) = 3x^4 + x^3 - 3x^2 + 7x + 5$

10. $h(x) = 7x^4 + 3x^3 - 2x^2 - x + 1$

Find all zeros of each function.

11. $h(x) = 2x^3 + 3x^2 - 65x + 84$

12. $p(x) = x^3 - 3x^2 + 9x - 7$

13. $h(x) = x^3 - 7x^2 + 17x - +5$

14. $q(x) = x^4 + 50x^2 + 49$

15. $g(x) = x^4 + 4x^3 - 3x^2 - 14x - 8$

16. $f(x) = x^4 - 6x^3 + 6x^2 + 24x - 40$

Write a polynomial function of least degree with integral coefficients that has the given zeros.

17. $-5, 3i$

18. $-2, 3 + i$

19. $-1, 4, 3i$

20. $2, 5, 1 + i$

21. CRAFTS Stephan has a set of plans to build a wooden box. He wants to reduce the volume of the box to 105 cubic inches. He would like to reduce the length of each dimension in the plan by the same amount. The plans call for the box to be 10 inches by 8 inches by 6 inches. Write and solve a polynomial equation to find out how much Stephan should take from each dimension.

NAME ______________________ DATE ____________ PERIOD ________

6-8 Skills Practice

Rational Zero Theorem

List all of the possible rational zeros of each function.

1. $n(x) = x^2 + 5x + 3$

2. $h(x) = x^2 - 2x - 5$

3. $w(x) = x^2 - 5x + 12$

4. $f(x) = 2x^2 + 5x + 3$

5. $q(x) = 6x^3 + x^2 - x + 2$

6. $g(x) = 9x^4 + 3x^3 + 3x^2 - x + 27$

Find all of the rational zeros of each function.

7. $f(x) = x^3 - 2x^2 + 5x - 4$

8. $g(x) = x^3 - 3x^2 - 4x + 12$

9. $p(x) = x^3 - x^2 + x - 1$

10. $z(x) = x^3 - 4x^2 + 6x - 4$

11. $h(x) = x^3 - x^2 + 4x - 4$

12. $g(x) = 3x^3 - 9x^2 - 10x - 8$

13. $g(x) = 2x^3 + 7x^2 - 7x - 12$

14. $h(x) = 2x^3 - 5x^2 - 4x + 3$

15. $p(x) = 3x^3 - 5x^2 - 14x - 4$

16. $q(x) = 3x^3 + 2x^2 + 27x + 18$

17. $q(x) = 3x^3 - 7x^2 + 4$

18. $f(x) = x^4 - 2x^3 - 13x^2 + 14x + 24$

19. $p(x) = x^4 - 5x^3 - 9x^2 - 25x - 70$

20. $n(x) = 16x^4 - 32x^3 - 13x^2 + 29x - 6$

Find all of the zeros of each function.

21. $f(x) = x^3 + 5x^2 + 11x + 15$

22. $q(x) = x^3 - 10x^2 + 18x - 4$

23. $m(x) = 6x^4 - 17x^3 + 8x^2 + 8x - 3$

24. $g(x) = x^4 + 4x^3 + 5x^2 + 4x + 4$

NAME ______________________ DATE ____________ PERIOD ____________

6-8 Practice

Rational Zero Theorem

List all of the possible rational zeros of each function.

1. $h(x) = x^3 - 5x^2 + 2x + 12$

2. $s(x) = x^4 - 8x^3 + 7x - 14$

3. $f(x) = 3x^5 - 5x^2 + x + 6$

4. $p(x) = 3x^2 + x + 7$

5. $g(x) = 5x^3 + x^2 - x + 8$

6. $q(x) = 6x^5 + x^3 - 3$

Find all of the rational zeros of each function.

7. $q(x) = x^3 + 3x^2 - 6x - 8$

8. $v(x) = x^3 - 9x^2 + 27x - 27$

9. $c(x) = x^3 - x^2 - 8x + 12$

10. $f(x) = x^4 - 49x^2$

11. $h(x) = x^3 - 7x^2 + 17x - 15$

12. $b(x) = x^3 + 6x + 20$

13. $f(x) = x^3 - 6x^2 + 4x - 24$

14. $g(x) = 2x^3 + 3x^2 - 4x - 4$

15. $h(x) = 2x^3 - 7x^2 - 21x + 54$

16. $z(x) = x^4 - 3x^3 + 5x^2 - 27x - 36$

17. $d(x) = x^4 + x^3 + 16$

18. $n(x) = x^4 - 2x^3 - 3$

19. $p(x) = 2x^4 - 7x^3 + 4x^2 + 7x - 6$

20. $q(x) = 6x^4 + -9x^3 + 40x^2 + 7x - 12$

Find all of the zeros of each function.

21. $f(x) = 2x^4 + 7x^3 - 2x^2 - 19x - 12$

22. $q(x) = x^4 - 4x^3 + x^2 + 16x - 20$

23. $h(x) = x^6 - 8x^3$

24. $g(x) = x^6 - 1$

25. TRAVEL The height of a box that Joan is shipping is 3 inches less than the width of the box. The length is 2 inches more than twice the width. The volume of the box is 1540 in^3. What are the dimensions of the box?

26. GEOMETRY The height of a square pyramid is 3 meters shorter than the side of its base. If the volume of the pyramid is 432 m^3, how tall is it? Use the formula $V = \frac{1}{3}Bh$.

NAME ______________________ DATE ____________ PERIOD ________

7-1 Skills Practice

Operations on Functions

Find $(f + g)(x)$, $(f - g)(x)$, $(f \cdot g)(x)$, and $\left(\frac{f}{g}\right)(x)$ for each $f(x)$ and $g(x)$.

1. $f(x) = x + 5$
$g(x) = x - 4$

2. $f(x) = 3x + 1$
$g(x) = 2x - 3$

3. $f(x) = x^2$
$g(x) = 4 - x$

4. $f(x) = 3x^2$
$g(x) = \frac{5}{x}$

For each pair of functions, find $f \circ g$ and $g \circ f$ if they exist.

5. $f = \{(0, 0), (4, -2)\}$
$g = \{(0, 4), (-2, 0), (5, 0)\}$

6. $f = \{(0, -3), (1, 2), (2, 2)\}$
$g = \{(-3, 1), (2, 0)\}$

7. $f = \{(-4, 3), (-1, 1), (2, 2)\}$
$g = \{(1, -4), (2, -1), (3, -1)\}$

8. $f = \{(6, 6), (-3, -3), (1, 3)\}$
$g = \{(-3, 6), (3, 6), (6, -3)\}$

Find $[g \circ h](x)$ and $[h \circ g](x)$ if they exist.

9. $g(x) = 2x$
$h(x) = x + 2$

10. $g(x) = -3x$
$h(x) = 4x - 1$

11. $g(x) = x - 6$
$h(x) = x + 6$

12. $g(x) = x - 3$
$h(x) = x^2$

13. $g(x) = 5x$
$h(x) = x^2 + x - 1$

14. $g(x) = x + 2$
$h(x) = 2x^2 - 3$

If $f(x) = 3x$, $g(x) = x + 4$, and $h(x) = x^2 - 1$, find each value.

15. $f[g(1)]$

16. $g[h(0)]$

17. $g[f(-1)]$

18. $h[f(5)]$

19. $g[h(-3)]$

20. $h[f(10)]$

21. $f[h(8)]$

22. $[f \circ (h \circ g)](1)$

23. $[f \circ (g \circ h)](-2)$

NAME ______________________________ DATE ____________ PERIOD ________

7-1 Practice

Operations on Functions

Find $(f + g)(x)$, $(f - g)(x)$, $(f \cdot g)(x)$, and $\left(\frac{f}{g}\right)(x)$ for each $f(x)$ and $g(x)$.

1. $f(x) = 2x + 1$
$g(x) = x - 3$

2. $f(x) = 8x^2$
$g(x) = \frac{1}{x^2}$

3. $f(x) = x^2 + 7x + 12$
$g(x) = x^2 - 9$

For each pair of functions, find $f \circ g$ and $g \circ f$, if they exist.

4. $f = \{(-9, -1), (-1, 0), (3, 4)\}$
$g = \{(0, -9), (-1, 3), (4, -1)\}$

5. $f = \{(-4, 3), (0, -2), (1, -2)\}$
$g = \{(-2, 0), (3, 1)\}$

6. $f = \{(-4, -5), (0, 3), (1, 6)\}$
$g = \{(6, 1), (-5, 0), (3, -4)\}$

7. $f = \{(0, -3), (1, -3), (6, 8)\}$
$g = \{(8, 2), (-3, 0), (-3, 1)\}$

Find $[g \circ h](x)$ and $[h \circ g](x)$, if they exist.

8. $g(x) = 3x$
$h(x) = x - 4$

9. $g(x) = -8x$
$h(x) = 2x + 3$

10. $g(x) = x + 6$
$h(x) = 3x^2$

11. $g(x) = x + 3$
$h(x) = 2x^2$

12. $g(x) = -2x$
$h(x) = x^2 + 3x + 2$

13. $g(x) = x - 2$
$h(x) = 3x^2 + 1$

If $f(x) = x^2$, $g(x) = 5x$, and $h(x) = x + 4$, find each value.

14. $f[g(1)]$

15. $g[h(-2)]$

16. $h[f(4)]$

17. $f[h(-9)]$

18. $h[g(-3)]$

19. $g[f(8)]$

20. BUSINESS The function $f(x) = 1000 - 0.01x^2$ models the manufacturing cost per item when x items are produced, and $g(x) = 150 - 0.001x^2$ models the service cost per item. Write a function $C(x)$ for the total manufacturing and service cost per item.

21. MEASUREMENT The formula $f = \frac{n}{12}$ converts inches n to feet f, and $m = \frac{f}{5280}$ converts feet to miles m. Write a composition of functions that converts inches to miles.

NAME ______________________ DATE ____________ PERIOD ________

7-2 Skills Practice

Inverse Functions and Relations

Find the inverse of each relation.

1. {(3, 1), (4, −3), (8, −3)}

2. {(−7, 1), (0, 5), (5, −1)}

3. {(−10, −2), (−7, 6), (−4, −2), (−4, 0)}

4. {(0, −9), (5, −3), (6, 6), (8, −3)}

5. {(−4, 12), (0, 7), (9, −1), (10, −5)}

6. {(−4, 1), (−4, 3), (0, −8), (8, −9)}

Find the inverse of each function. Then graph the function and its inverse.

7. $y = 4$

8. $f(x) = 3x$

9. $f(x) = x + 2$

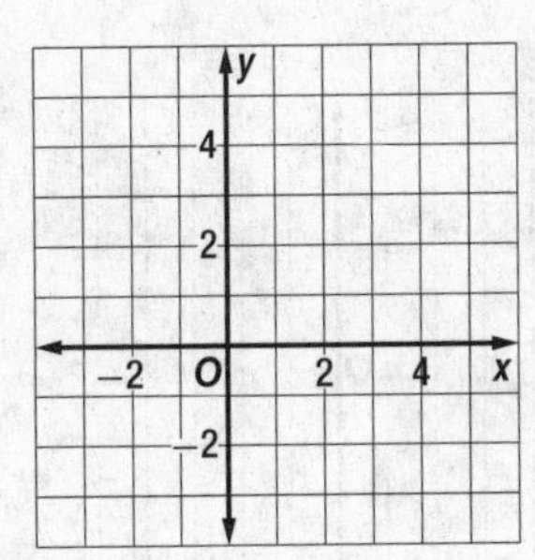

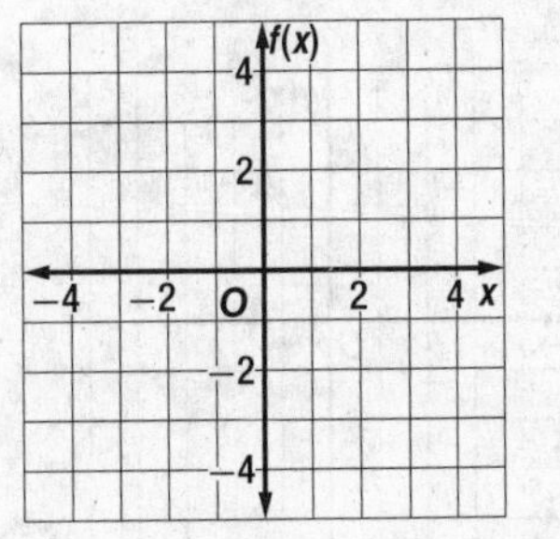

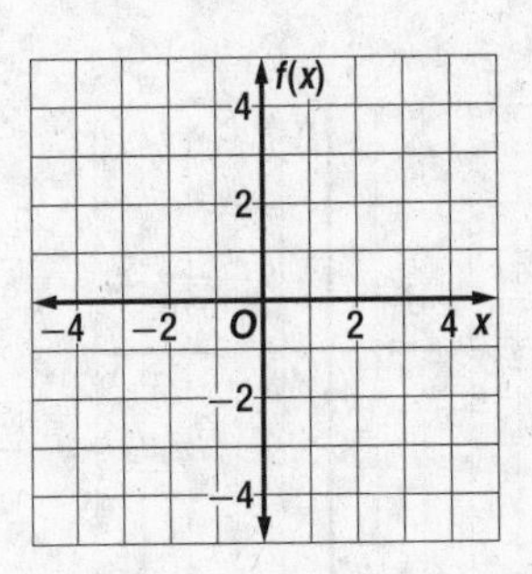

10. $g(x) = 2x - 1$

11. $h(x) = \frac{1}{4}x$

12. $y = \frac{2}{3}x + 2$

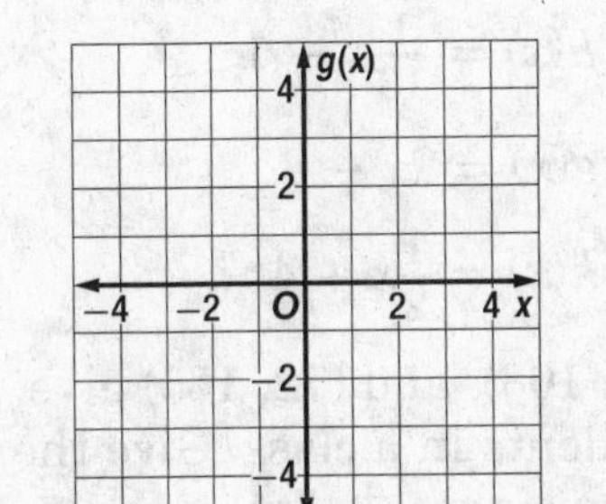

Determine whether each pair of functions are inverse functions. Write *yes* or *no*.

13. $f(x) = x - 1$
$g(x) = 1 - x$

14. $f(x) = 2x + 3$
$g(x) = \frac{1}{2}(x - 3)$

15. $f(x) = 5x - 5$
$g(x) = \frac{1}{5}x + 1$

16. $f(x) = 2x$
$g(x) = \frac{1}{2}x$

17. $h(x) = 6x - 2$
$g(x) = \frac{1}{6}x + 3$

18. $f(x) = 8x - 10$
$g(x) = \frac{1}{8}x + \frac{5}{4}$

NAME ________________ DATE ________ PERIOD ________

7-2 Practice

Inverse Functions and Relations

Find the inverse of each relation.

1. {(0, 3), (4, 2), (5, −6)}

2. {(−5, 1), (−5, −1), (−5, 8)}

3. {(−3, −7), (0, −1), (5, 9), (7, 13)}

4. {(8, −2), (10, 5), (12, 6), (14, 7)}

5. {(−5, −4), (1, 2), (3, 4), (7, 8)}

6. {(−3, 9), (−2, 4), (0, 0), (1, 1)}

Find the inverse of each function. Then graph the function and its inverse.

7. $f(x) = \frac{3}{4}x$

8. $g(x) = 3 + x$

9. $y = 3x - 2$

f(x) O x

g(x) O x

y O x

Determine whether each pair of functions are inverse functions. Write *yes* or *no*.

10. $f(x) = x + 6$
$g(x) = x - 6$

11. $f(x) = -4x + 1$
$g(x) = \frac{1}{4}(1 - x)$

12. $g(x) = 13x - 13$
$h(x) = \frac{1}{13}x - 1$

13. $f(x) = 2x$
$g(x) = -2x$

14. $f(x) = \frac{6}{7}x$
$g(x) = \frac{7}{6}x$

15. $g(x) = 2x - 8$
$h(x) = \frac{1}{2}x + 4$

16. MEASUREMENT The points (63, 121), (71, 180), (67, 140), (65, 108), and (72, 165) give the weight in pounds as a function of height in inches for 5 students in a class. Give the points for these students that represent height as a function of weight.

17. REMODELING The Clearys are replacing the flooring in their 15 foot by 18 foot kitchen. The new flooring costs $17.99 per square yard. The formula $f(x) = 9x$ converts square yards to square feet.

a. Find the inverse $f^{-1}(x)$. What is the significance of $f^{-1}(x)$ for the Clearys?

b. What will the new flooring cost the Clearys?

NAME ______________________ DATE ____________ PERIOD ____________

7-3 Skills Practice

Square Root Functions and Inequalities

Graph each function. State the domain and range of each function.

1. $y = \sqrt{2x}$

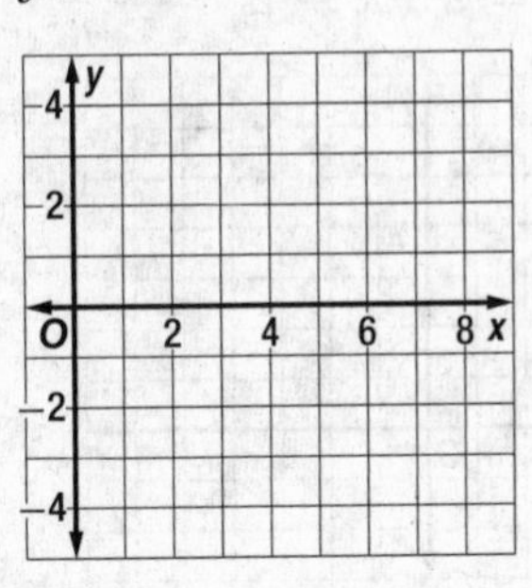

2. $y = -\sqrt{3x}$

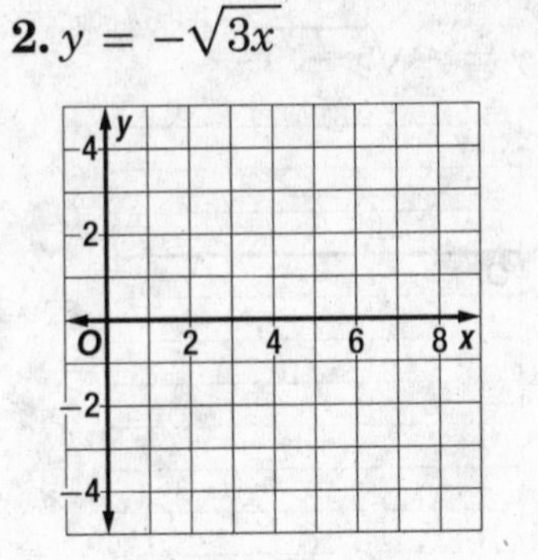

3. $y = 2\sqrt{x}$

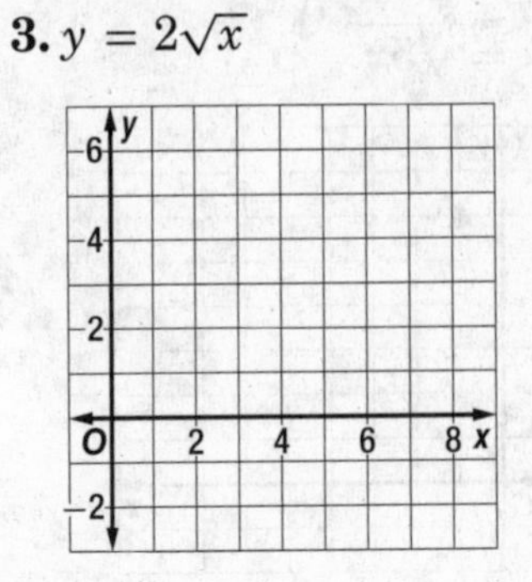

4. $y = \sqrt{x + 3}$

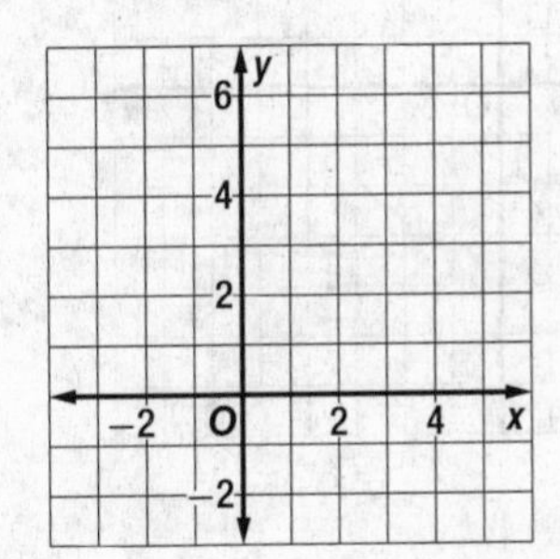

5. $y = -\sqrt{2x - 5}$

6. $y = \sqrt{x + 4} - 2$

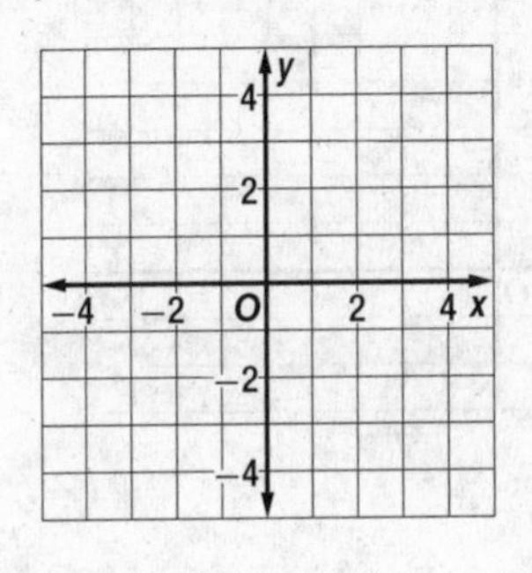

Graph each inequality.

7. $f(x) < \sqrt{4x}$

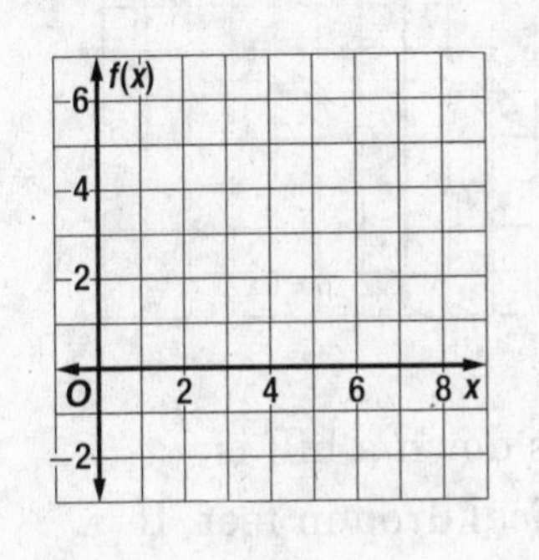

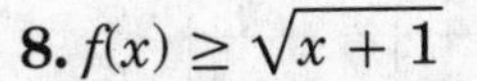

8. $f(x) \geq \sqrt{x + 1}$

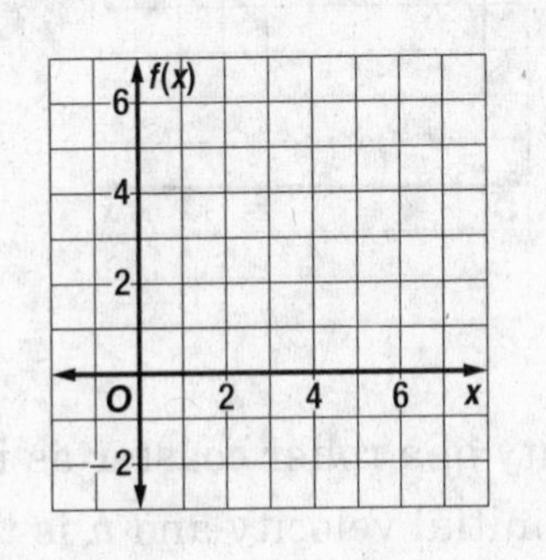

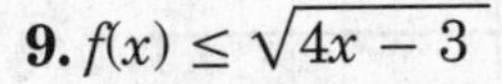

9. $f(x) \leq \sqrt{4x - 3}$

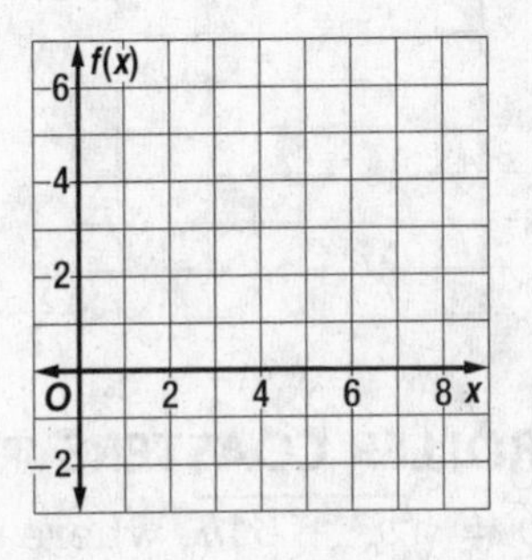

7-3 Practice

Square Root Functions and Inequalities

Graph each function. State the domain and range.

1. $y = \sqrt{5x}$

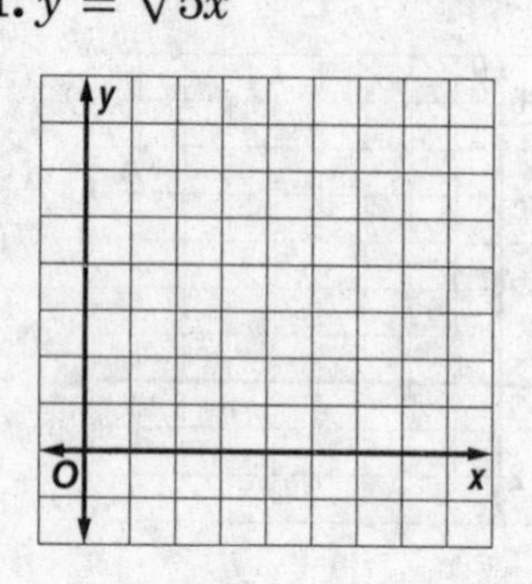

2. $y = -\sqrt{x - 1}$

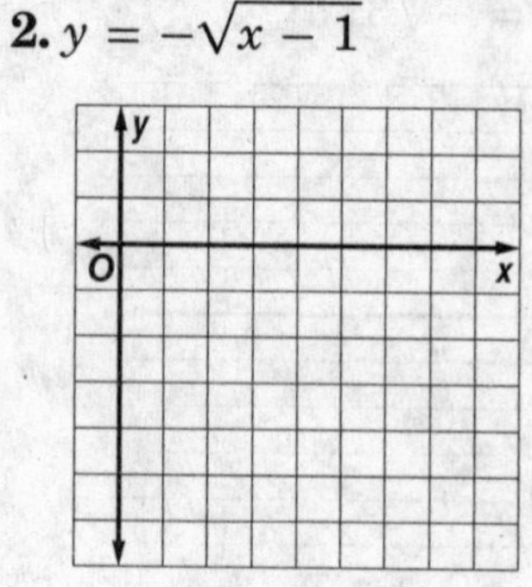

3. $y = 2\sqrt{x + 2}$

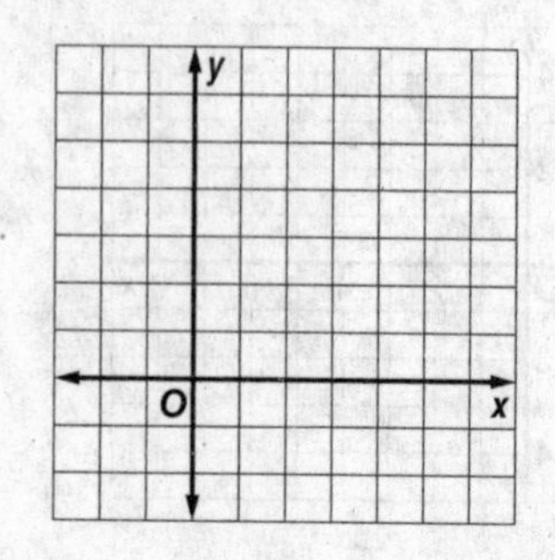

4. $y = \sqrt{3x - 4}$

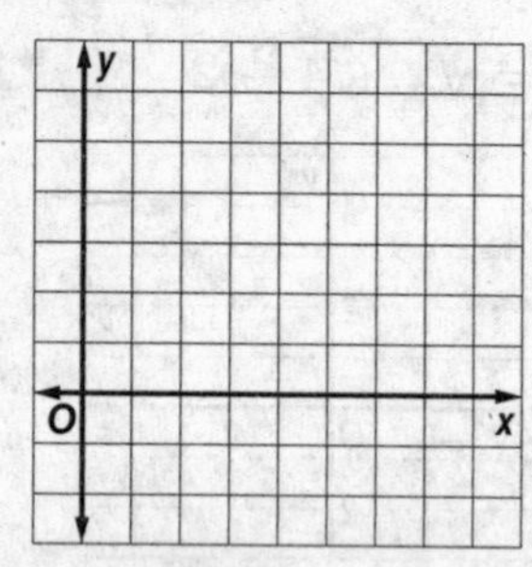

5. $y = \sqrt{x + 7} - 4$

6. $y = 1 - \sqrt{2x + 3}$

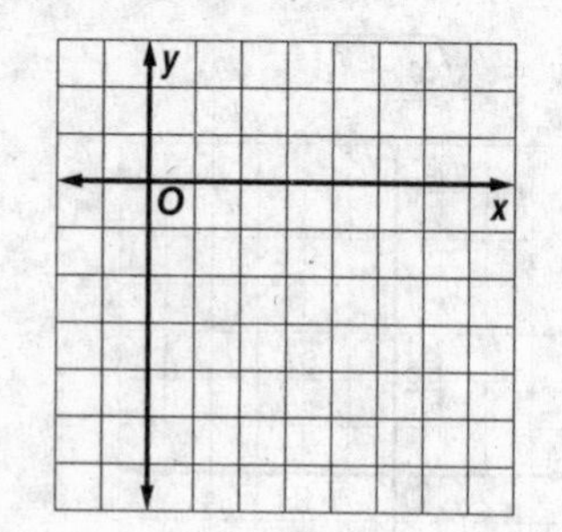

Graph each inequality.

7. $y \geq -\sqrt{6x}$

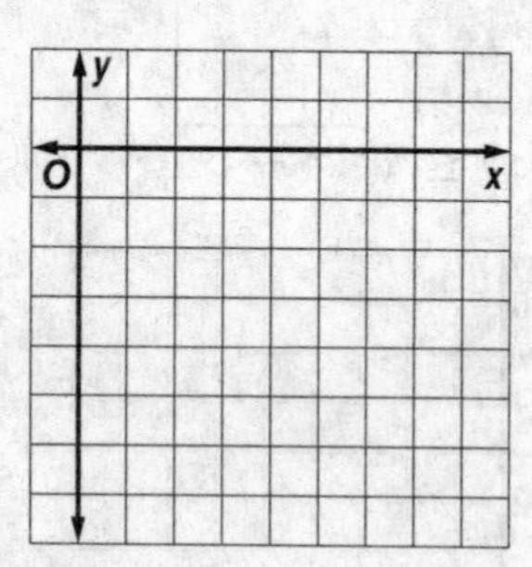

8. $y \leq \sqrt{x - 5} + 3$

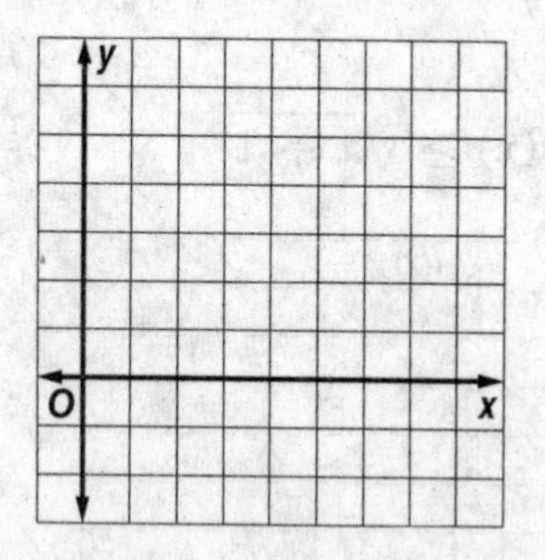

9. $y > -2\sqrt{3x + 2}$

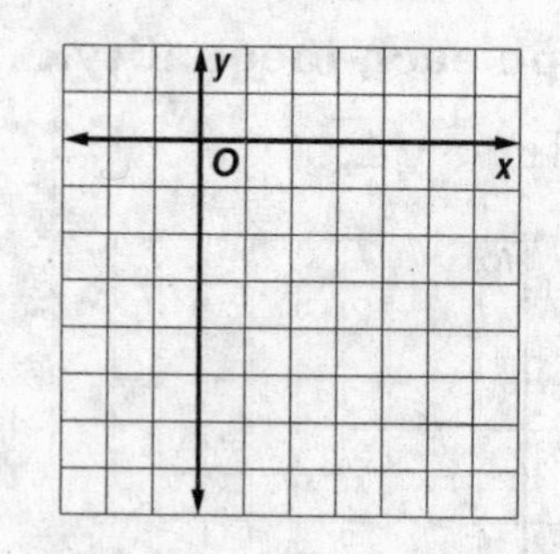

10. ROLLER COASTERS The velocity of a roller coaster as it moves down a hill is $v = \sqrt{v_0^2 + 64h}$, where v_0 is the initial velocity and h is the vertical drop in feet. If $v = 70$ feet per second and $v_0 = 8$ feet per second, find h.

11. WEIGHT Use the formula $d = \sqrt{\frac{3960^2 W_E}{W_s}} - 3960$, which relates distance from Earth d in miles to weight. If an astronaut's weight on Earth W_E is 148 pounds and in space W_s is 115 pounds, how far from Earth is the astronaut?

NAME ______________________ DATE ____________ PERIOD ________

7-4 Skills Practice

*n*th Roots

Use a calculator to approximate each value to three decimal places.

1. $\sqrt{230}$

2. $\sqrt{38}$

3. $-\sqrt{152}$

4. $\sqrt{5.6}$

5. $\sqrt[3]{88}$

6. $\sqrt[3]{-222}$

7. $-\sqrt[4]{0.34}$

8. $\sqrt[5]{500}$

Simplify.

9. $\pm\sqrt{81}$

10. $\sqrt{144}$

11. $\sqrt{(-5)^2}$

12. $\sqrt{-5^2}$

13. $\sqrt{0.36}$

14. $-\sqrt{\frac{4}{9}}$

15. $\sqrt[3]{-8}$

16. $-\sqrt[3]{27}$

17. $\sqrt[3]{0.064}$

18. $\sqrt[5]{32}$

19. $\sqrt[4]{81}$

20. $\sqrt{y^2}$

21. $\sqrt[3]{125c^3}$

22. $\sqrt{64x^6}$

23. $\sqrt[3]{-27a^6}$

24. $\sqrt{m^8p^4}$

25. $-\sqrt{100p^4t^2}$

26. $\sqrt[4]{16w^4v^8}$

27. $\sqrt{(-3c)^4}$

28. $\sqrt{(a+b)^2}$

NAME ______________________________ DATE ____________ PERIOD ________

7-4 Practice

nth Roots

Simplify.

1. $\sqrt{0.81}$
2. $-\sqrt{324}$
3. $-\sqrt[4]{256}$
4. $\sqrt[6]{64}$
5. $\sqrt[3]{-64}$
6. $\sqrt[3]{0.512}$
7. $\sqrt[5]{-243}$
8. $-\sqrt[4]{1296}$
9. $\sqrt[5]{\frac{-1024}{243}}$
10. $\sqrt[5]{243x^{10}}$
11. $\sqrt{14a^2}$
12. $\sqrt{-(14a)^2}$
13. $\sqrt{49m^2t^8}$
14. $\sqrt{\frac{16m^2}{25}}$
15. $\sqrt[3]{-64r^2w^{15}}$
16. $\sqrt{(2x)^8}$
17. $-\sqrt[4]{625s^8}$
18. $\sqrt[3]{216p^3q^9}$
19. $\sqrt{676x^4y^6}$
20. $\sqrt[3]{-27x^9y^{12}}$
21. $-\sqrt{144m^8n^6}$
22. $\sqrt[5]{-32x^5y^{10}}$
23. $\sqrt[6]{(m+4)^6}$
24. $\sqrt[3]{(2x+1)^3}$
25. $-\sqrt{49a^{10}b^{16}}$
26. $\sqrt[4]{(x-5)^8}$
27. $\sqrt[3]{343d^6}$
28. $\sqrt{x^2+10x+25}$

Use a calculator to approximate each value to three decimal places.

29. $\sqrt{7.8}$
30. $-\sqrt{89}$
31. $\sqrt[3]{25}$
32. $\sqrt[3]{-4}$
33. $\sqrt[4]{1.1}$
34. $\sqrt[5]{-0.1}$
35. $\sqrt[6]{5555}$
36. $\sqrt[4]{(0.94)^2}$

37. **RADIANT TEMPERATURE** Thermal sensors measure an object's *radiant* temperature, which is the amount of energy radiated by the object. The *internal* temperature of an object is called its *kinetic* temperature. The formula $T_r = T_k\sqrt[4]{e}$ relates an object's radiant temperature T_r to its kinetic temperature T_k. The variable e in the formula is a measure of how well the object radiates energy. If an object's kinetic temperature is 30°C and $e = 0.94$, what is the object's radiant temperature to the nearest tenth of a degree?

38. **HERO'S FORMULA** Salvatore is buying fertilizer for his triangular garden. He knows the lengths of all three sides, so he is using Hero's formula to find the area. Hero's formula states that the area of a triangle is $\sqrt{s(s-a)(s-b)(s-c)}$, where a, b, and c are the lengths of the sides of the triangle and s is half the perimeter of the triangle. If the lengths of the sides of Salvatore's garden are 15 feet, 17 feet, and 20 feet, what is the area of the garden? Round your answer to the nearest whole number.

NAME ______________ DATE ________ PERIOD ________

7-5 Skills Practice

Operations with Radical Expressions

Simplify.

1. $\sqrt{24}$

2. $\sqrt{75}$

3. $\sqrt[3]{16}$

4. $-\sqrt[4]{48}$

5. $4\sqrt{50x^5}$

6. $\sqrt[4]{64a^4b^4}$

7. $\sqrt[3]{-8d^2f^5}$

8. $\sqrt{\frac{25}{36}r^2t}$

9. $-\sqrt{\frac{3}{7}}$

10. $\sqrt[3]{\frac{2}{9}}$

11. $\sqrt{\frac{2g^3}{5z}}$

12. $(3\sqrt{3})(5\sqrt{3})$

13. $(4\sqrt{12})(3\sqrt{20})$

14. $\sqrt{2} + \sqrt{8} + \sqrt{50}$

15. $\sqrt{12} - 2\sqrt{3} + \sqrt{108}$

16. $8\sqrt{5} - \sqrt{45} - \sqrt{80}$

17. $2\sqrt{48} - \sqrt{75} - \sqrt{12}$

18. $(2 + \sqrt{3})(6 - \sqrt{2})$

19. $(1 - \sqrt{5})(1 + \sqrt{5})$

20. $(3 - \sqrt{7})(5 + \sqrt{2})$

21. $(\sqrt{2} - \sqrt{6})^2$

22. $\frac{3}{7 - \sqrt{2}}$

23. $\frac{4}{3 + \sqrt{2}}$

24. $\frac{5}{8 - \sqrt{6}}$

NAME ______________________ DATE ____________ PERIOD ____________

7-5 Practice

Operations with Radical Expressions

Simplify.

1. $\sqrt{540}$

2. $\sqrt[3]{-432}$

3. $\sqrt[3]{128}$

4. $-\sqrt[4]{405}$

5. $\sqrt[3]{-5000}$

6. $\sqrt[5]{-1215}$

7. $\sqrt[3]{125t^6w^2}$

8. $\sqrt[4]{48v^8z^{13}}$

9. $\sqrt[3]{8g^3k^8}$

10. $\sqrt{45x^3y^8}$

11. $\sqrt{\frac{11}{9}}$

12. $\sqrt[3]{\frac{216}{24}}$

13. $\sqrt{\frac{1}{128}c^4d^7}$

14. $\sqrt{\frac{9a^5}{64b^4}}$

15. $\sqrt[4]{\frac{8}{9a^3}}$

16. $(3\sqrt{15})(-4\sqrt{45})$

17. $(2\sqrt{24})(7\sqrt{18})$

18. $\sqrt{810} + \sqrt{240} - \sqrt{250}$

19. $6\sqrt{20} + 8\sqrt{5} - 5\sqrt{45}$

20. $8\sqrt{48} - 6\sqrt{75} + 7\sqrt{80}$

21. $(3\sqrt{2} + 2\sqrt{3})^2$

22. $(3 - \sqrt{7})^2$

23. $(\sqrt{5} - \sqrt{6})(\sqrt{5} + \sqrt{2})$

24. $(\sqrt{2} + \sqrt{10})(\sqrt{2} - \sqrt{10})$

25. $(1 + \sqrt{6})(5 - \sqrt{7})$

26. $(\sqrt{3} + 4\sqrt{7})^2$

27. $(\sqrt{108} - 6\sqrt{3})^2$

28. $\frac{\sqrt{3}}{\sqrt{5} - 2}$

29. $\frac{6}{\sqrt{2} - 1}$

30. $\frac{5 + \sqrt{3}}{4 + \sqrt{3}}$

31. $\frac{3 + \sqrt{2}}{2 - \sqrt{2}}$

32. $\frac{3 + \sqrt{6}}{5 - \sqrt{24}}$

33. $\frac{3 + \sqrt{x}}{2 - \sqrt{x}}$

34. BRAKING The formula $s = 2\sqrt{5\ell}$ estimates the speed s in miles per hour of a car when it leaves skid marks ℓ feet long. Use the formula to write a simplified expression for s if $\ell = 85$. Then evaluate s to the nearest mile per hour.

35. PYTHAGOREAN THEOREM The measures of the legs of a right triangle can be represented by the expressions $6x^2y$ and $9x^2y$. Use the Pythagorean Theorem to find a simplified expression for the measure of the hypotenuse.

NAME ______________________ DATE ____________ PERIOD ________

7-6 Skills Practice

Rational Exponents

Write each expression in radical form, or write each radical in exponential form.

1. $3^{\frac{1}{6}}$

2. $8^{\frac{1}{5}}$

3. $\sqrt{51}$

4. $\sqrt[4]{15^3}$

5. $12^{\frac{2}{3}}$

6. $\sqrt[3]{37}$

7. $(c^3)^{\frac{3}{5}}$

8. $\sqrt[3]{6xy^2}$

Evaluate each expression.

9. $32^{\frac{1}{5}}$

10. $81^{\frac{1}{4}}$

11. $27^{\frac{1}{3}}$

12. $4^{\frac{1}{2}}$

13. $16^{\frac{3}{2}}$

14. $(-243)^{\frac{4}{5}}$

15. $27^{\frac{1}{3}} \cdot 27^{\frac{5}{3}}$

16. $\left(\frac{4}{9}\right)^{\frac{3}{2}}$

Simplify each expression.

17. $c^{\frac{12}{5}} \cdot c^{\frac{3}{5}}$

18. $m^{\frac{2}{9}} \cdot m^{\frac{16}{9}}$

19. $\left(q^{\frac{1}{2}}\right)^3$

20. $p^{\frac{1}{5}} \cdot p^{\frac{1}{2}}$

21. $x^{\frac{6}{11}} \cdot x^{\frac{4}{11}}$

22. $\dfrac{x^{\frac{2}{3}}}{x^{\frac{1}{4}}}$

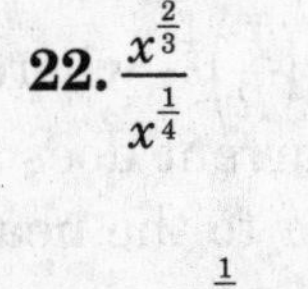

23. $\dfrac{y^{\frac{1}{2}}}{y^{\frac{1}{4}}}$

24. $\dfrac{n^{\frac{1}{3}}}{n^{\frac{1}{6}} \cdot n^{\frac{1}{2}}}$

25. $\sqrt[12]{64}$

26. $\sqrt[8]{49a^8b^2}$

7-6 Practice

Rational Exponents

Write each expression in radical form, or write each radical in exponential form.

1. $5^{\frac{1}{3}}$

2. $6^{\frac{2}{5}}$

3. $m^{\frac{4}{7}}$

4. $(n^3)^{\frac{2}{5}}$

5. $\sqrt{79}$

6. $\sqrt[4]{153}$

7. $\sqrt[3]{27m^6n^4}$

8. $\sqrt[5]{2a^{10}b}$

Evaluate each expression.

9. $81^{\frac{1}{4}}$

10. $1024^{-\frac{1}{5}}$

11. $8^{-\frac{5}{3}}$

12. $-256^{-\frac{3}{4}}$

13. $(-64)^{-\frac{2}{3}}$

14. $27^{\frac{1}{3}} \cdot 27^{\frac{4}{3}}$

15. $\left(\frac{125}{216}\right)^{-\frac{2}{3}}$

16. $\frac{64^{\frac{2}{3}}}{343^{\frac{2}{3}}}$

17. $\left(25^{\frac{1}{2}}\right)\left(-64^{-\frac{1}{3}}\right)$

Simplify each expression.

18. $g^{\frac{4}{7}} \cdot g^{\frac{3}{7}}$

19. $s^{\frac{3}{4}} \cdot s^{\frac{13}{4}}$

20. $\left(u^{-\frac{1}{3}}\right)^{-\frac{4}{5}}$

21. $y^{-\frac{1}{2}}$

22. $b^{-\frac{3}{5}}$

23. $\frac{q^{\frac{3}{5}}}{q^{\frac{2}{5}}}$

24. $\frac{t^{\frac{2}{3}}}{5t^{\frac{1}{2}} \cdot t^{-\frac{3}{4}}}$

25. $\frac{2z^{\frac{1}{2}}}{z^{\frac{1}{2}} - 1}$

26. $\sqrt[10]{8^5}$

27. $\sqrt{12} \cdot \sqrt[5]{12^3}$

28. $\sqrt[4]{6} \cdot 3\sqrt[4]{6}$

29. $\frac{a}{\sqrt{3b}}$

30. ELECTRICITY The amount of current in amps I that an appliance uses can be calculated using the formula $I = \left(\frac{P}{R}\right)^{\frac{1}{2}}$, where P is the power in watts and R is the resistance in ohms. How much current does an appliance use if $P = 500$ watts and $R = 10$ ohms? Round your answer to the nearest tenth.

31. BUSINESS A company that produces DVDs uses the formula $C = 88n^{\frac{1}{3}} + 330$ to calculate the cost C in dollars of producing n DVDs per day. What is the company's cost to produce 150 DVDs per day? Round your answer to the nearest dollar.

NAME ______________________ DATE ____________ PERIOD ____________

7-7 Skills Practice

Solving Radical Equations and Inequalities

Solve each equation.

1. $\sqrt{x} = 5$

2. $\sqrt{x} + 3 = 7$

3. $5\sqrt{j} = 1$

4. $v^{\frac{1}{2}} + 1 = 0$

5. $18 - 3y^{\frac{1}{2}} = 25$

6. $\sqrt[3]{2w} = 4$

7. $\sqrt{b - 5} = 4$

8. $\sqrt{3n + 1} = 5$

9. $\sqrt[3]{3r - 6} = 3$

10. $2 + \sqrt{3p + 7} = 6$

11. $\sqrt{k - 4} - 1 = 5$

12. $(2d + 3)^{\frac{1}{3}} = 2$

13. $(t - 3)^{\frac{1}{3}} = 2$

14. $4 - (1 - 7u)^{\frac{1}{3}} = 0$

15. $\sqrt{3z - 2} = \sqrt{z - 4}$

16. $\sqrt{g + 1} = \sqrt{2g - 7}$

Solve each inequality.

17. $4\sqrt{x + 1} \geq 12$

18. $5 + \sqrt{c - 3} \leq 6$

19. $-2 + \sqrt{3x + 3} < 7$

20. $-\sqrt{2a + 4} \geq -6$

21. $2\sqrt{4r - 3} > 10$

22. $4 - \sqrt{3x + 1} > 3$

23. $\sqrt{y + 4} - 3 \geq 3$

24. $-3\sqrt{11r + 3} \geq -15$

NAME ______________________ DATE ____________ PERIOD ________

7-7 Practice

Solving Radical Equations and Inequalities

Solve each equation.

1. $\sqrt{x} = 8$

2. $4 - \sqrt{x} = 3$

3. $\sqrt{2p} + 3 = 10$

4. $4\sqrt{3h} - 2 = 0$

5. $c^{\frac{1}{2}} + 6 = 9$

6. $18 + 7h^{\frac{1}{2}} = 12$

7. $\sqrt[3]{d + 2} = 7$

8. $\sqrt[5]{w - 7} = 1$

9. $6 + \sqrt[3]{q - 4} = 9$

10. $\sqrt[4]{y - 9} + 4 = 0$

11. $\sqrt{2m - 6} - 16 = 0$

12. $\sqrt[3]{4m + 1} - 2 = 2$

13. $\sqrt{8n - 5} - 1 = 2$

14. $\sqrt{1 - 4t} - 8 = -6$

15. $\sqrt{2t - 5} - 3 = 3$

16. $(7v - 2)^{\frac{1}{4}} + 12 = 7$

17. $(3g + 1)^{\frac{1}{2}} - 6 = 4$

18. $(6u - 5)^{\frac{1}{3}} + 2 = -3$

19. $\sqrt{2d - 5} = \sqrt{d - 1}$

20. $\sqrt{4r - 6} = \sqrt{r}$

21. $\sqrt{6x - 4} = \sqrt{2x + 10}$

22. $\sqrt{2x + 5} = \sqrt{2x + 1}$

Solve each inequality.

23. $3\sqrt{a} \geq 12$

24. $\sqrt{z + 5} + 4 \leq 13$

25. $8 + \sqrt{2q} \leq 5$

26. $\sqrt{2a - 3} < 5$

27. $9 - \sqrt{c + 4} \leq 6$

28. $\sqrt{x - 1} < 2$

29. STATISTICS Statisticians use the formula $\sigma = \sqrt{v}$ to calculate a standard deviation σ, where v is the variance of a data set. Find the variance when the standard deviation is 15.

30. GRAVITATION Helena drops a ball from 25 feet above a lake. The formula $t = \frac{1}{4}\sqrt{25 - h}$ describes the time t in seconds that the ball is h feet above the water. How many feet above the water will the ball be after 1 second?

NAME ______________________ DATE ____________ PERIOD ________

8-1 Skills Practice

Graphing Exponential Functions

Graph each function. State the function's domain and range.

1. $y = 3(2)^x$

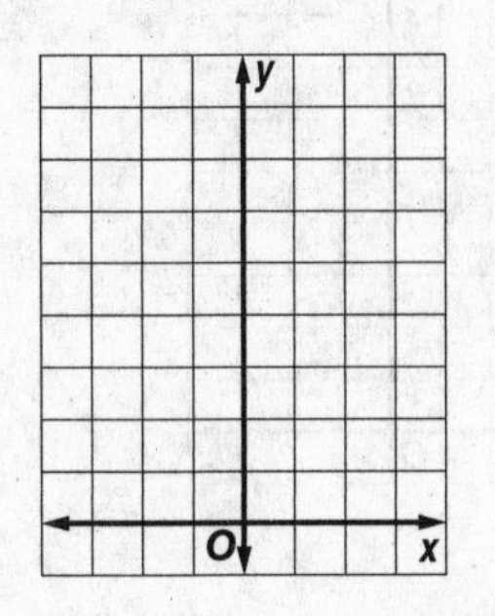

2. $y = 2\left(\frac{1}{2}\right)^x$

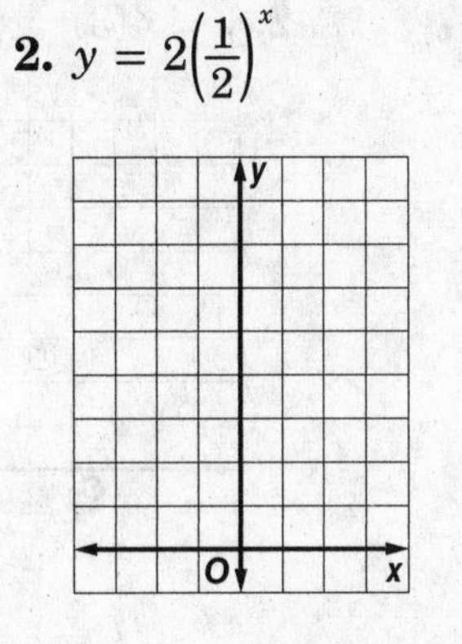

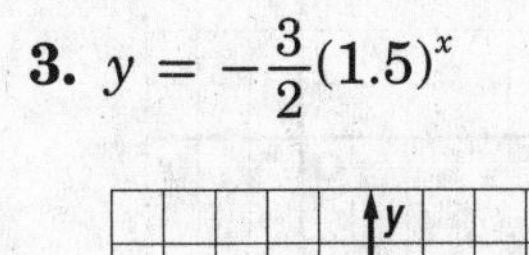

3. $y = -\frac{3}{2}(1.5)^x$

4. $y = 3\left(\frac{1}{3}\right)^x$

For each graph $f(x)$ is the parent function and $g(x)$ is a transformation of $f(x)$. Use the graph to determine $g(x)$.

5. $f(x) = 4^x$

6. $f(x) = \left(\frac{1}{5}\right)^x$

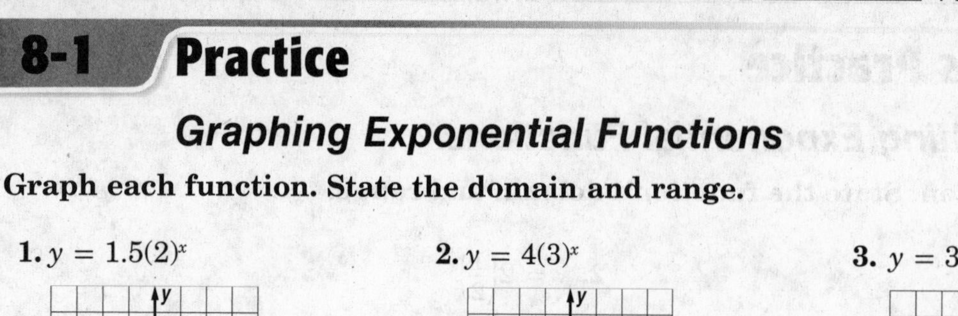

8-1 Practice

Graphing Exponential Functions

Graph each function. State the domain and range.

1. $y = 1.5(2)^x$

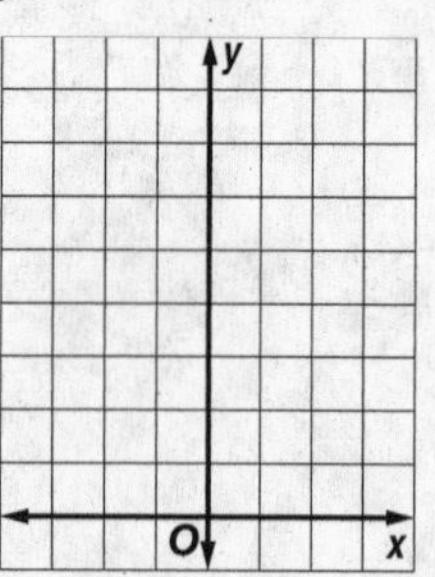

2. $y = 4(3)^x$

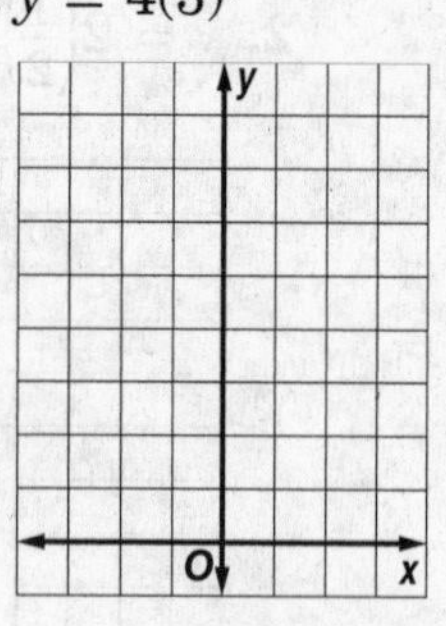

3. $y = 3(0.5)^x$

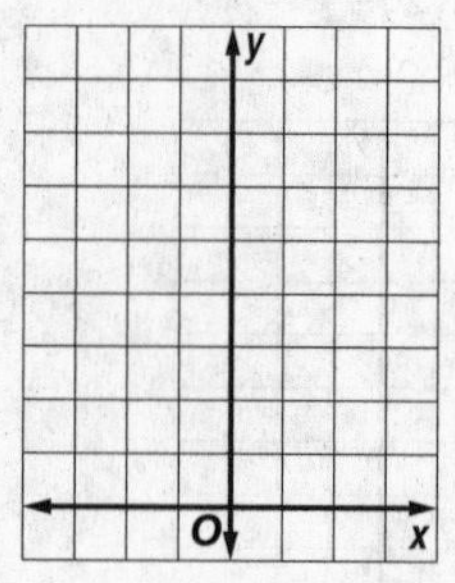

4. $y = 5\left(\frac{1}{2}\right)^x - 8$

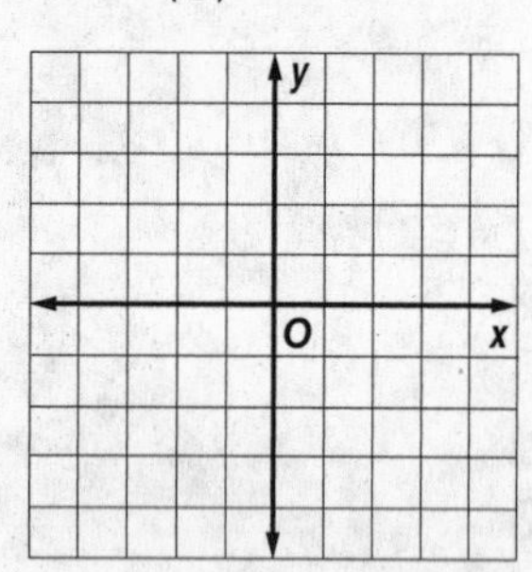

5. $y = -2\left(\frac{1}{4}\right)^{x-3}$

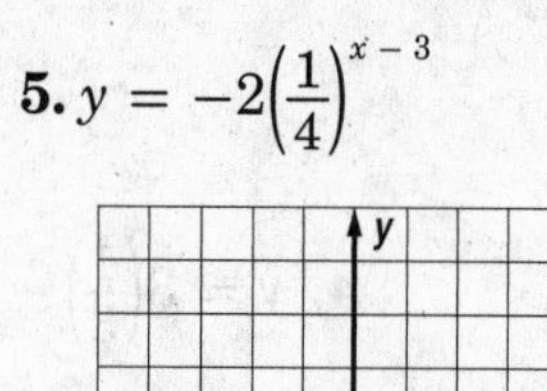

6. $y = \frac{1}{2}(3)^{x+4} - 5$

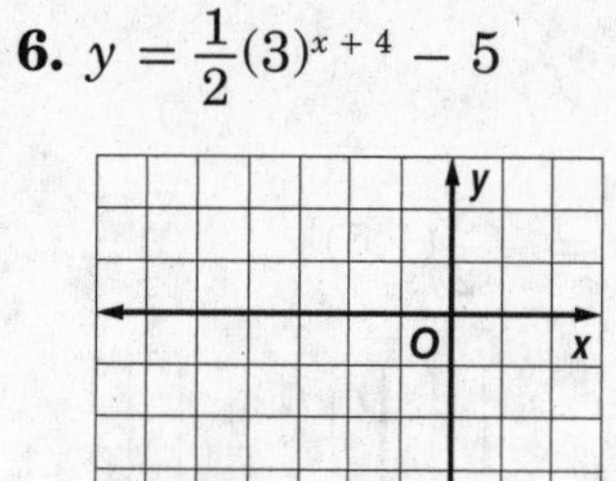

7. BIOLOGY The initial number of bacteria in a culture is 12,000. The culture doubles each day.

a. Write an exponential function to model the population y of bacteria after x days.

b. How many bacteria are there after 6 days?

8. EDUCATION A college with a graduating class of 4000 students in the year 2008 predicts that its graduating class will grow 5% per year. Write an exponential function to model the number of students y in the graduating class t years after 2008.

NAME ______________________ DATE ____________ PERIOD ____________

8-2 Skills Practice

Solving Exponential Equations and Inequalities

Solve each equation.

1. $25^{2x+3} = 25^{5x-9}$

2. $9^{8x-4} = 81^{3x+6}$

3. $4^{x-5} = 16^{2x-31}$

4. $4^{3x-3} = 8^{4x-4}$

5. $9^{-x+5} = 27^{6x-10}$

6. $125^{3x-4} = 25^{4x+2}$

Solve each inequality.

7. $\left(\frac{1}{36}\right)^{6x-3} > 6^{3x-9}$

8. $64^{4x-8} < 256^{2x+6}$

9. $\left(\frac{1}{27}\right)^{3x+13} \leq 9^{5x-\frac{1}{2}}$

10. $\left(\frac{1}{9}\right)^{2x+7} \leq 27^{6x-12}$

11. $\left(\frac{1}{8}\right)^{-2x-6} > \left(\frac{1}{32}\right)^{-x+11}$

12. $9^{9x+1} < \left(\frac{1}{243}\right)^{-3x+5}$

Write an exponential function whose graph passes through the given points.

13. (0, 3) and (3, 375)

14. (0, −1) and (6, −64)

15. (0, 7) and (−2, 28)

16. $\left(0, \frac{1}{2}\right)$ and (2, 40.5)

17. (0, 15) and (1, 12)

18. (0, −6) and (−4, −1536)

19. $\left(0, \frac{1}{3}\right)$ and (3, 9)

20. (0, 1) and (6, 4096)

21. (0, −2) and (−1, −4)

NAME ______________________ DATE ____________ PERIOD ________

8-2 Practice

Solving Exponential Equations and Inequalities

Solve each equation.

1. $4^{x+35} = 64^{x-3}$

2. $\left(\frac{1}{64}\right)^{0.5x-3} = 8^{9x-2}$

3. $3^{x-4} = 9^{x+28}$

4. $\left(\frac{1}{4}\right)^{2x+2} = 64^{x-1}$

5. $\left(\frac{1}{2}\right)^{x-3} = 16^{3x+1}$

6. $3^{6x-2} = \left(\frac{1}{9}\right)^{x+1}$

Write an exponential function for the graph that passes through the given points.

7. (0, 5) and (4, 3125)

8. (0, 8) and (4, 2048)

9. $\left(0, \frac{3}{4}\right)$ and (2, 36.75)

10. (0, –0.2) and (–3, –3.125)

11. (0, 15) and $\left(2, \frac{15}{16}\right)$

12. (0, 0.7) and $\left(\frac{1}{2}, 3.5\right)$

Solve each inequality.

13. $400 > \left(\frac{1}{20}\right)^{7x+8}$

14. $10^{2x+7} \geq 1000^{x}$

15. $\left(\frac{1}{16}\right)^{3x-4} \leq 64^{x-1}$

16. $\left(\frac{1}{8}\right)^{x-6} < 4^{4x+5}$

17. $\left(\frac{1}{36}\right)^{x+8} \leq 216^{x-3}$

18. $128^{x+3} < \left(\frac{1}{1024}\right)^{2x}$

19. At time t, there are 216^{t+18} bacteria of type A and 36^{2t+8} bacteria of type B organisms in a sample. When will the number of each type of bacteria be equal?

8-3 Skills Practice

Logarithms and Logarithmic Functions

Write each equation in exponential form.

1. $\log_3 243 = 5$

2. $\log_4 64 = 3$

3. $\log_9 3 = \frac{1}{2}$

4. $\log_5 \frac{1}{25} = -2$

Write each equation in logarithmic form.

5. $2^3 = 8$

6. $3^2 = 9$

7. $8^{-2} = \frac{1}{64}$

8. $\left(\frac{1}{3}\right)^2 = \frac{1}{9}$

Evaluate each expression.

9. $\log_5 25$

10. $\log_9 3$

11. $\log_{10} 1000$

12. $\log_{125} 5$

13. $\log_4 \frac{1}{64}$

14. $\log_5 \frac{1}{625}$

15. $\log_8 512$

16. $\log_{27} \frac{1}{3}$

Graph each function.

17. $f(x) = \log_3 (x + 1) - 4$

f(x)
O x

18. $f(x) = -\log_5 x + 2.5$

f(x)
O x

8-3 Practice

Logarithms and Logarithmic Functions

Write each equation in exponential form.

1. $\log_6 216 = 3$

2. $\log_2 64 = 6$

3. $\log_3 \frac{1}{81} = -4$

4. $\log_{10} 0.00001 = -5$

5. $\log_{25} 5 = \frac{1}{2}$

6. $\log_{32} 8 = \frac{3}{5}$

Write each equation in logarithmic form.

7. $5^3 = 125$

8. $7^0 = 1$

9. $3^4 = 81$

10. $3^{-4} = \frac{1}{81}$

11. $\left(\frac{1}{4}\right)^3 = \frac{1}{64}$

12. $7776^{\frac{1}{5}} = 6$

Evaluate each expression.

13. $\log_3 81$

14. $\log_{10} 0.0001$

15. $\log_2 \frac{1}{16}$

16. $\log_{\frac{1}{3}} 27$

17. $\log_9 1$

18. $\log_8 4$

19. $\log_7 \frac{1}{49}$

20. $\log_6 6^4$

Graph each function.

21. $f(x) = \log_2 (x - 2)$

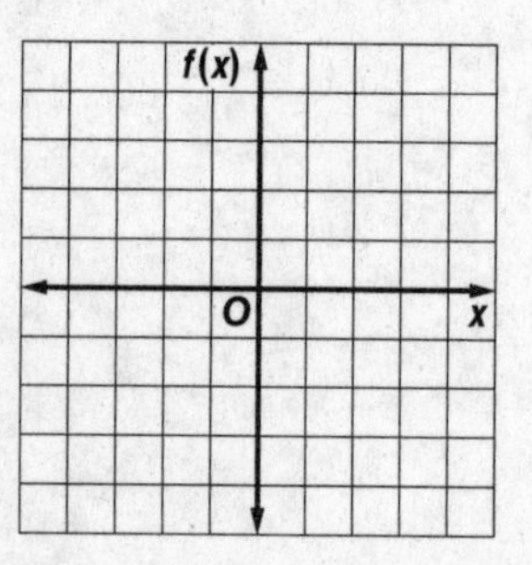

22. $f(x) = -2 \log_4 x$

f(x)
O
x

23. SOUND An equation for loudness, in decibels, is $L = 10 \log_{10} R$, where R is the relative intensity of the sound. Sounds that reach levels of 120 decibels or more are painful to humans. What is the relative intensity of 120 decibels?

24. INVESTING Maria invests $1000 in a savings account that pays 4% interest compounded annually. The value of the account A at the end of five years can be determined from the equation $\log_{10} A = \log_{10}[1000(1 + 0.04)^5]$. Write this equation in exponential form.

NAME ______________________ DATE __________ PERIOD __________

8-4 Skills Practice

Solving Logarithmic Equations and Inequalities

Solve each equation.

1. $3x = \log_6 216$

2. $x - 4 = \log_3 243$

3. $\log_4 (4x - 20) = 5$

4. $\log_9 (3 - x) = \log_9 (5x - 15)$

5. $\log_{81} (x + 20) = \log_{81} (6x)$

6. $\log_9 (3x^2) = \log_9 (2x + 1)$

7. $\log_4 (x - 1) = \log_4 (12)$

8. $\log_7 (5 - x) = \log_7 (5)$

9. $\log_x (5x) = 2$

Solve each inequality.

10. $\log_5 (-3x) < 1$

11. $\log_6 x > \log_6 (4 - x)$

12. $\log_{10} (x - 3) < 2$

13. $\log_2 (x - 5) > \log_2 (3)$

14. $\log_7 (8x + 5) > \log_7 (6x - 18)$

15. $\log_9 (3x - 3) < 1.5$

16. $\log_{10} (2x - 2) < \log_{10} (7 - x)$

17. $\log_9 (x - 1) > \log_9 (2x)$

18. $\log_{16} x \geq 0.5$

19. $\log_3\left(\frac{x - 3}{4} + 5\right) > \log_3 (x + 2)$

20. $\log_5 (3x) < \log_5 (2x - 1)$

21. $\log_3 (7 - x) \leq \log_3 (x + 19)$

NAME ______________________ DATE ____________ PERIOD ________

8-4 Practice

Solving Logarithmic Equations and Inequalities

Solve each equation.

1. $x + 5 = \log_4 256$

2. $3x - 5 = \log_2 1024$

3. $\log_3 (4x - 17) = 5$

4. $\log_5 (3 - x) = 5$

5. $\log_{13} (x^2 - 4) = \log_{13} 3x$

6. $\log_3 (x - 5) = \log_3 (3x - 25)$

Solve each inequality

7. $\log_8 (-6x) < 1$

8. $\log_9 (x + 2) > \log_9 (6 - 3x)$

9. $\log_{11} (x + 7) < 1$

10. $\log_{81} x \leq 0.75$

11. $\log_2 (x + 6) < \log_2 17$

12. $\log_{12} (2x - 1) > \log_{12} (5x - 16)$

13. $\log_9 (2x - 1) < 0.5$

14. $\log_{10} (x - 5) > \log_{10} 2x$

15. $\log_3 (x + 12) > \log_3 2x$

16. $\log_3 (0.3x + 5) > \log_3 (x - 2)$

17. $\log_2 (x + 3) < \log_2 (1 - 3x)$

18. $\log_6 (3 - x) \leq \log_6 (x - 1)$

19. WILDLIFE An ecologist discovered that the population of a certain endangered species has been doubling every 12 years. When the population reaches 20 times the current level, it may no longer be endangered. Write the logarithmic expression that gives the number of years it will take for the population to reach that level.

NAME ______________________ DATE __________ PERIOD __________

8-5 Skills Practice

Properties of Logarithms

Use $\log_2 3 \approx 1.5850$ and $\log_2 5 \approx 2.3219$ to approximate the value of each expression.

1. $\log_2 25$

2. $\log_2 27$

3. $\log_2 \frac{3}{5}$

4. $\log_2 \frac{5}{3}$

5. $\log_2 15$

6. $\log_2 45$

7. $\log_2 75$

8. $\log_2 0.6$

9. $\log_2 \frac{1}{3}$

10. $\log_2 \frac{9}{5}$

Solve each equation. Check your solutions.

11. $\log_{10} 27 = 3 \log_{10} x$

12. $3 \log_7 4 = 2 \log_7 b$

13. $\log_4 5 + \log_4 x = \log_4 60$

14. $\log_6 2c + \log_6 8 = \log_6 80$

15. $\log_5 y - \log_5 8 = \log_5 1$

16. $\log_2 q - \log_2 3 = \log_2 7$

17. $\log_9 4 + 2 \log_9 5 = \log_9 w$

18. $3 \log_8 2 - \log_8 4 = \log_8 b$

19. $\log_{10} x + \log_{10} (3x - 5) = \log_{10} 2$

20. $\log_4 x + \log_4 (2x - 3) = \log_4 2$

21. $\log_3 d + \log_3 3 = 3$

22. $\log_{10} y - \log_{10} (2 - y) = 0$

23. $\log_2 r + 2 \log_2 5 = 0$

24. $\log_2 (x + 4) - \log_2 (x - 3) = 3$

25. $\log_4 (n + 1) - \log_4 (n - 2) = 1$

26. $\log_5 10 + \log_5 12 = 3 \log_5 2 + \log_5 a$

NAME ______________________ DATE ____________ PERIOD ________

8-5 Practice

Properties of Logarithms

Use $\log_{10} 5 \approx 0.6990$ and $\log_{10} 7 \approx 0.8451$ to approximate the value of each expression.

1. $\log_{10} 35$ **2.** $\log_{10} 25$ **3.** $\log_{10} \frac{7}{5}$ **4.** $\log_{10} \frac{5}{7}$

5. $\log_{10} 245$ **6.** $\log_{10} 175$ **7.** $\log_{10} 0.2$ **8.** $\log_{10} \frac{25}{7}$

Solve each equation. Check your solutions.

9. $\log_7 n = \frac{2}{3} \log_7 8$

10. $\log_{10} u = \frac{3}{2} \log_{10} 4$

11. $\log_6 x + \log_6 9 = \log_6 54$

12. $\log_8 48 - \log_8 w = \log_8 4$

13. $\log_9 (3u + 14) - \log_9 5 = \log_9 2u$

14. $4 \log_2 x + \log_2 5 = \log_2 405$

15. $\log_3 y = -\log_3 16 + \frac{1}{3} \log_3 64$

16. $\log_2 d = 5 \log_2 2 - \log_2 8$

17. $\log_{10} (3m - 5) + \log_{10} m = \log_{10} 2$

18. $\log_{10} (b + 3) + \log_{10} b = \log_{10} 4$

19. $\log_8 (t + 10) - \log_8 (t - 1) = \log_8 12$

20. $\log_3 (a + 3) + \log_3 (a + 2) = \log_3 6$

21. $\log_{10} (r + 4) - \log_{10} r = \log_{10} (r + 1)$

22. $\log_4 (x^2 - 4) - \log_4 (x + 2) = \log_4 1$

23. $\log_{10} 4 + \log_{10} w = 2$

24. $\log_8 (n - 3) + \log_8 (n + 4) = 1$

25. $3 \log_5 (x^2 + 9) - 6 = 0$

26. $\log_{16} (9x + 5) - \log_{16} (x^2 - 1) = \frac{1}{2}$

27. $\log_6 (2x - 5) + 1 = \log_6 (7x + 10)$

28. $\log_2 (5y + 2) - 1 = \log_2 (1 - 2y)$

29. $\log_{10} (c^2 - 1) - 2 = \log_{10} (c + 1)$

30. $\log_7 x + 2 \log_7 x - \log_7 3 = \log_7 72$

31. SOUND Recall that the loudness L of a sound in decibels is given by $L = 10 \log_{10} R$, where R is the sound's relative intensity. If the intensity of a certain sound is tripled, by how many decibels does the sound increase?

32. EARTHQUAKES An earthquake rated at 3.5 on the Richter scale is felt by many people, and an earthquake rated at 4.5 may cause local damage. The Richter scale magnitude reading m is given by $m = \log_{10} x$, where x represents the amplitude of the seismic wave causing ground motion. How many times greater is the amplitude of an earthquake that measures 4.5 on the Richter scale than one that measures 3.5?

NAME ______________________ DATE __________ PERIOD __________

8-6 Skills Practice

Common Logarithms

Use a calculator to evaluate each expression to the nearest ten-thousandth.

1. $\log 6$

2. $\log 15$

3. $\log 1.1$

4. $\log 0.3$

Solve each equation or inequality. Round to the nearest ten-thousandth.

5. $3^x > 243$

6. $16^v \leq \frac{1}{4}$

7. $8^p = 50$

8. $7^y = 15$

9. $5^{3b} = 106$

10. $4^{5k} = 37$

11. $12^{7p} = 120$

12. $9^{2m} = 27$

13. $3^{r-5} = 4.1$

14. $8^{y+4} > 15$

15. $7.6^{d+3} = 57.2$

16. $0.5^{t-8} = 16.3$

17. $42^{x^2} = 84$

18. $5^{x^2+1} = 10$

Express each logarithm in terms of common logarithms. Then approximate its value to the nearest ten-thousandth.

19. $\log_3 7$

20. $\log_5 66$

21. $\log_2 35$

22. $\log_6 10$

23. Use the formula $\text{pH} = -\log [H+]$ to find the pH of each substance given its concentration of hydrogen ions. Round to the nearest tenth.

a. gastric juices: $[H+] = 1.0 \times 10^{-1}$ mole per liter

b. tomato juice: $[H+] = 7.94 \times 10^{-5}$ mole per liter

c. blood: $[H+] = 3.98 \times 10^{-8}$ mole per liter

d. toothpaste: $[H+] = 1.26 \times 10^{-10}$ mole per liter

NAME ______________________ DATE ____________ PERIOD ____________

8-6 Practice

Common Logarithms

Use a calculator to evaluate each expression to the nearest ten-thousandth.

1. log 101

2. log 2.2

3. log 0.05

Use the formula pH = −log [H+] to find the pH of each substance given its concentration of hydrogen ions. Round to the nearest tenth.

4. milk: $[H+] = 2.51 \times 10^{-7}$ mole per liter

5. acid rain: $[H+] = 2.51 \times 10^{-6}$ mole per liter

6. black coffee: $[H+] = 1.0 \times 10^{-5}$ mole per liter

7. milk of magnesia: $[H+] = 3.16 \times 10^{-11}$ mole per liter

Solve each equation or inequality. Round to the nearest ten-thousandth.

8. $2^x < 25$

9. $5^a = 120$

10. $6^z = 45.6$

11. $9^m \geq 100$

12. $3.5^x = 47.9$

13. $8.2^y = 64.5$

14. $2^{b+1} \leq 7.31$

15. $4^{2x} = 27$

16. $2^{a-4} = 82.1$

17. $9^{z-2} > 38$

18. $5^{w+3} = 17$

19. $30^{x^2} = 50$

20. $5^{x^2-3} = 72$

21. $4^{2x} = 9^{x+1}$

22. $2^{n+1} = 5^{2n-1}$

Express each logarithm in terms of common logarithms. Then approximate its value to the nearest ten-thousandth.

23. $\log_5 12$

24. $\log_8 32$

25. $\log_{11} 9$

26. $\log_2 18$

27. $\log_9 6$

28. $\log_7 \sqrt{8}$

29. HORTICULTURE Siberian irises flourish when the concentration of hydrogen ions [H+] in the soil is not less than 1.58×10^{-8} mole per liter. What is the pH of the soil in which these irises will flourish?

30. ACIDITY The pH of vinegar is 2.9 and the pH of milk is 6.6. Approximately how many times greater is the hydrogen ion concentration of vinegar than of milk?

31. BIOLOGY There are initially 1000 bacteria in a culture. The number of bacteria doubles each hour. The number of bacteria N present after t hours is $N = 1000(2)^t$. How long will it take the culture to increase to 50,000 bacteria?

32. SOUND An equation for loudness L in decibels is given by $L = 10 \log R$, where R is the sound's relative intensity. An air-raid siren can reach 150 decibels and jet engine noise can reach 120 decibels. How many times greater is the relative intensity of the air-raid siren than that of the jet engine noise?

NAME ______________________ DATE ____________ PERIOD ________

8-7 Skills Practice

Base e and Natural Logarithms

Write an equivalent exponential or logarithmic equation.

1. $e^x = 3$

2. $e^4 = 8x$

3. $\ln 15 = x$

4. $\ln x \approx 0.6931$

5. $e^4 = x - 3$

6. $\ln 5.34 = 2x$

Write each as a single logarithm.

7. $3 \ln 3 - \ln 9$

8. $4 \ln 16 - \ln 256$

9. $2 \ln x + 2 \ln 4$

10. $3 \ln 4 + 3 \ln 3$

Solve each equation or inequality. Round to the nearest ten-thousandth.

11. $e^x \geq 5$

12. $e^x < 3.2$

13. $2e^x - 1 = 11$

14. $5e^x + 3 = 18$

15. $e^{3x} = 30$

16. $e^{-4x} > 10$

17. $e^{5x} + 4 > 34$

18. $1 - 2e^{2x} = -19$

19. $\ln 3x = 2$

20. $\ln 8x = 3$

21. $\ln (x - 2) = 2$

22. $\ln (x + 3) = 1$

23. $\ln (x + 3) = 4$

24. $\ln x + \ln 2x = 2$

NAME ______________________ DATE ____________ PERIOD ________

8-7 Practice

Base e and Natural Logarithms

Write an equivalent exponential or logarithmic equation.

1. $\ln 50 = x$

2. $\ln 36 = 2x$

3. $\ln 6 \approx 1.7918$

4. $\ln 9.3 \approx 2.2300$

5. $e^x = 8$

6. $e^5 = 10x$

7. $e^{-x} = 4$

8. $e^2 = x + 1$

Solve each equation or inequality. Round to four decimal places.

9. $e^x < 9$

10. $e^{-x} = 31$

11. $e^x = 1.1$

12. $e^x = 5.8$

13. $2e^x - 3 = 1$

14. $5e^x + 1 \geq 7$

15. $4 + e^x = 19$

16. $-3e^x + 10 < 8$

17. $e^{3x} = 8$

18. $e^{-4x} = 5$

19. $e^{0.5x} = 6$

20. $2e^{5x} = 24$

21. $e^{2x} + 1 = 55$

22. $e^{3x} - 5 = 32$

23. $9 + e^{2x} = 10$

24. $e^{-3x} + 7 \geq 15$

25. $\ln 4x = 3$

26. $\ln (-2x) = 7$

27. $\ln 2.5x = 10$

28. $\ln (x - 6) = 1$

29. $\ln (x + 2) = 3$

30. $\ln (x + 3) = 5$

31. $\ln 3x + \ln 2x = 9$

32. $\ln 5x + \ln x = 7$

33. INVESTING Sarita deposits $1000 in an account paying 3.4% annual interest compounded continuously. Use the formula for continuously compounded interest, $A = Pe^{rt}$, where P is the principal, r is the annual interest rate, and t is the time in years.

a. What is the balance in Sarita's account after 5 years?

b. How long will it take the balance in Sarita's account to reach $2000?

34. RADIOACTIVE DECAY The amount of a radioactive substance y that remains after t years is given by the equation $y = ae^{kt}$, where a is the initial amount present and k is the decay constant for the radioactive substance. If $a = 100$, $y = 50$, and $k = -0.035$, find t.

8-8 Skills Practice

Using Exponential and Logarithmic Functions

1. **FISHING** In an over-fished area, the catch of a certain fish is decreasing exponentially. Use $k = 0.084$ to determine how long will it take for the catch to reach half of its current the amount?

2. **POPULATION** A current census shows that the population of a city is 3.5 million. Using the formula $P = ae^{rt}$, find the expected population of the city in 30 years if the growth rate r of the population is 1.5%, a represents the current population in millions, and t represents the time in years.

3. **POPULATION** The population P in thousands of a city can be modeled by the equation $P = 80e^{0.015t}$, where t is the time in years. In how many years will the population of the city be 120,000?

4. **BACTERIA** How many days will it take a culture of bacteria to increase from 2000 to 50,000? Use $k = 0.657$.

5. **NUCLEAR POWER** The element plutonium-239 is highly radioactive. Nuclear reactors can produce and also use this element. The heat that plutonium-239 emits has helped to power equipment on the moon. If the half-life of plutonium-239 is 24,360 years, what is the value of k for this element?

6. **DEPRECIATION** A Global Positioning Satellite (GPS) system uses satellite information to locate ground position. Abu's surveying firm bought a GPS system for $12,500. The GPS is now worth $8600. How long ago did Abu buy the GPS system? Use $k = 0.062$.

7. **LOGISTIC GROWTH** The population of a certain habitat follows the function

$$p(t) = \frac{105{,}000}{1 + 2.7e^{-0.0981t}}.$$

a. What is the maximum population of this habitat?

b. When does the population reach 100,000? Round to the nearest hundredth.

NAME ______________________ DATE __________ PERIOD __________

8-8 Practice

Using Exponential and Logarithmic Functions

1. **BACTERIA** How many hours will it take a culture of bacteria to increase from 20 to 2000? Use $k = 0.614$.

2. **RADIOACTIVE DECAY** A radioactive substance has a half-life of 32 years. Find the constant k in the decay formula for the substance.

3. **RADIOACTIVE DECAY** Cobalt, an element used to make alloys, has several isotopes. One of these, cobalt 60, is radioactive and has a half-life of 5.7 years. Cobalt 60 is used to trace the path of nonradioactive substances in a system. What is the value of k for cobalt 60?

4. **WHALES** Modern whales appeared 5–10 million years ago. The vertebrae of a whale discovered by paleontologists contain roughly 0.25% as much carbon-14 as they would have contained when the whale was alive. How long ago did the whale die? Use $k = 0.00012$.

5. **POPULATION** The population of rabbits in an area is modeled by the growth equation $P(t) = 8e^{0.26t}$, where P is in thousands and t is in years. How long will it take for the population to reach 25,000?

6. **RADIOACTIVE DECAY** A radioactive element decays exponentially. The decay model is given by the formula $A = A_0e^{-0.04463t}$. A is the amount present after t days and A_0 is the amount present initially. Assume you are starting with 50g. How much of the element remains after 10 days? 30 days?

7. **POPULATION** A population is growing continuously at a rate of 3%. If the population is now 5 million, what will it be in 17 years' time?

8. **BACTERIA** A certain bacteria is growing exponentially according to the model $y = 80e^{kt}$. Using $k = 0.071$, find how many hours it will take for the bacteria reach a population of 10,000 cells?

9. **LOGISTIC GROWTH** The population of a certain habitat follows the function

$$P(t) = \frac{16{,}300}{(1 + 17.5e^{-0.065t})}.$$

a. What is the maximum population?

b. When does the population reach 16,200?

NAME ______________________ DATE ____________ PERIOD ________

9-1 Skills Practice

Multiplying and Dividing Rational Expressions

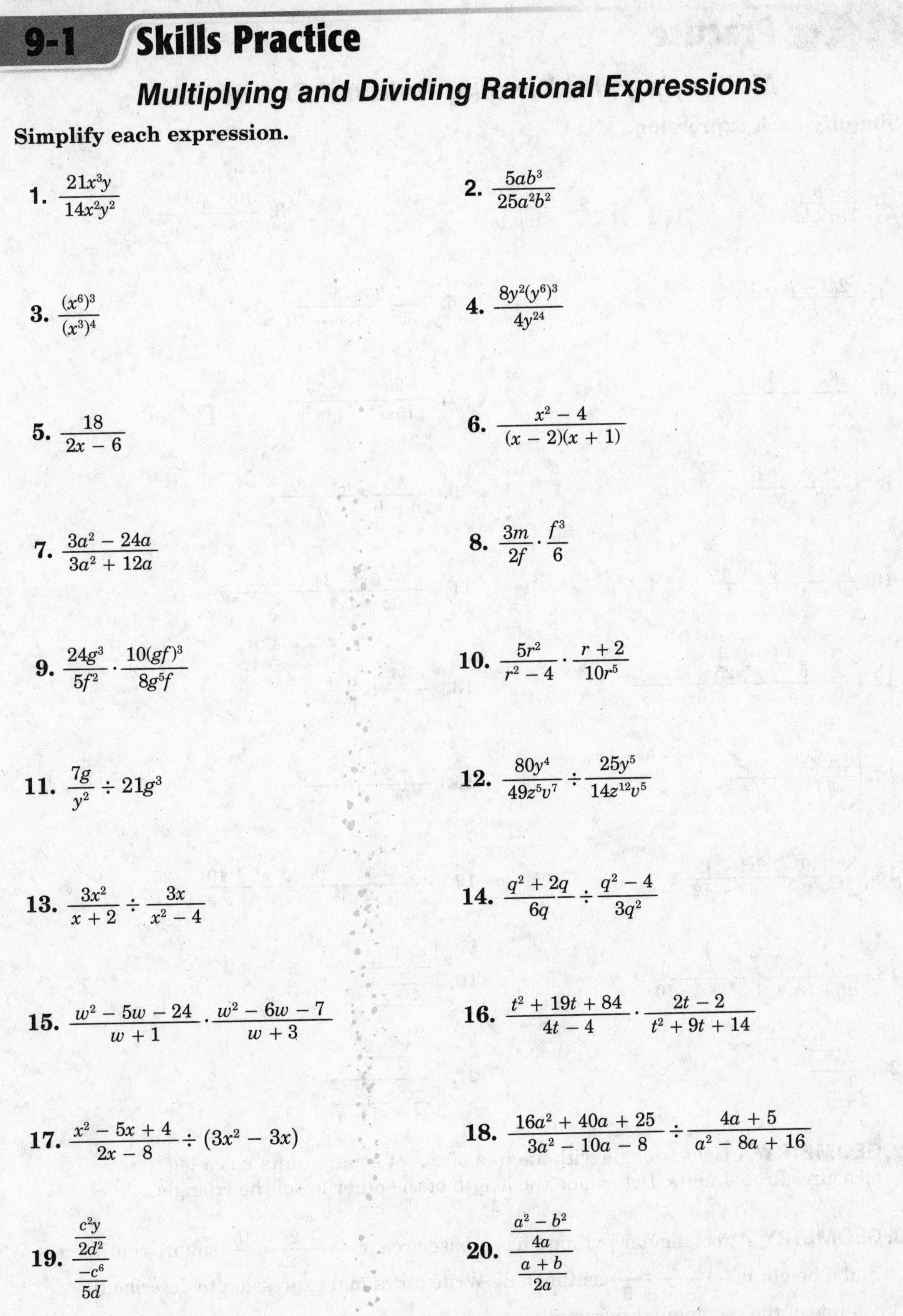

Simplify each expression.

1. $\frac{21x^3y}{14x^2y^2}$

2. $\frac{5ab^3}{25a^2b^2}$

3. $\frac{(x^6)^3}{(x^3)^4}$

4. $\frac{8y^2(y^6)^3}{4y^{24}}$

5. $\frac{18}{2x - 6}$

6. $\frac{x^2 - 4}{(x - 2)(x + 1)}$

7. $\frac{3a^2 - 24a}{3a^2 + 12a}$

8. $\frac{3m}{2f} \cdot \frac{f^3}{6}$

9. $\frac{24g^3}{5f^2} \cdot \frac{10(gf)^3}{8g^5f}$

10. $\frac{5r^2}{r^2 - 4} \cdot \frac{r + 2}{10r^5}$

11. $\frac{7g}{y^2} \div 21g^3$

12. $\frac{80y^4}{49z^5v^7} \div \frac{25y^5}{14z^{12}v^5}$

13. $\frac{3x^2}{x + 2} \div \frac{3x}{x^2 - 4}$

14. $\frac{q^2 + 2q}{6q} \div \frac{q^2 - 4}{3q^2}$

15. $\frac{w^2 - 5w - 24}{w + 1} \cdot \frac{w^2 - 6w - 7}{w + 3}$

16. $\frac{t^2 + 19t + 84}{4t - 4} \cdot \frac{2t - 2}{t^2 + 9t + 14}$

17. $\frac{x^2 - 5x + 4}{2x - 8} \div (3x^2 - 3x)$

18. $\frac{16a^2 + 40a + 25}{3a^2 - 10a - 8} \div \frac{4a + 5}{a^2 - 8a + 16}$

19. $\frac{\frac{c^2y}{2d^2}}{\frac{-c^6}{5d}}$

20. $\frac{\frac{a^2 - b^2}{4a}}{\frac{a + b}{2a}}$

NAME ______________________ DATE ____________ PERIOD ____________

9-1 Practice

Multiplying and Dividing Rational Expressions

Simplify each expression.

1. $\frac{9a^2b^3}{27a^4b^4c}$

2. $\frac{(2m^3n^2)^3}{-18m^5n^4}$

3. $\frac{10y^2 + 15y}{35y^2 - 5y}$

4. $\frac{2k^2 - k - 15}{k^2 - 9}$

5. $\frac{25 - v^2}{3v^2 - 13v - 10}$

6. $\frac{x^4 + x^3 - 2x^2}{x^4 - x^3}$

7. $\frac{-2u^3y}{15xz^5} \cdot \frac{25x^3}{14u^2y^2}$

8. $\frac{a + y}{6} \cdot \frac{4}{y + a}$

9. $\frac{n^5}{n - 6} \cdot \frac{n^2 - 6n}{n^8}$

10. $\frac{a - y}{w + n} \cdot \frac{w^2 - n^2}{y - a}$

11. $\frac{x^2 - 5x - 24}{6x + 2x^2} \cdot \frac{5x^2}{8 - x}$

12. $\frac{x - 5}{10x - 2} \cdot \frac{25x^2 - 1}{x^2 - 10x + 25}$

13. $\frac{a^5y^3}{wy^7} \div \frac{a^3w^2}{w^5y^2}$

14. $\left(\frac{2xy}{w^2}\right)^3 \div \frac{24x^2}{w^5}$

15. $\frac{x + y}{6} \div \frac{x^2 - y^2}{3}$

16. $\frac{3x + 6}{x^2 - 9} \div \frac{6x^2 + 12x}{4x + 12}$

17. $\frac{2s^2 - 7s - 15}{(s + 4)^2} \div \frac{s^2 - 10s + 25}{s + 4}$

18. $\frac{9 - a^2}{a^2 + 5a + 6} \div \frac{2a - 6}{5a + 10}$

19. $\frac{\frac{2x + 1}{x}}{\frac{4 - x}{x}}$

20. $\frac{\frac{x^2 - 9}{4}}{\frac{3 - x}{8}}$

21. $\frac{\frac{x^3 + 2^3}{x^2 - 2x}}{\frac{(x + 2)^3}{x^2 + 4x + 4}}$

22. GEOMETRY A right triangle with an area of $x^2 - 4$ square units has a leg that measures $2x + 4$ units. Determine the length of the other leg of the triangle.

23. GEOMETRY A rectangular pyramid has a base area of $\frac{x^2 + 3x - 10}{2x}$ square centimeters and a height of $\frac{x^2 - 3x}{x^2 - 5x + 6}$ centimeters. Write a rational expression to describe the volume of the rectangular pyramid.

NAME ______________________ DATE ____________ PERIOD ____________

9-2 Skills Practice

Adding and Subtracting Rational Expressions

Find the LCM of each set of polynomials.

1. $12c, 6c^2d$

2. $18a^3bc^2, 24b^2c^2$

3. $2x - 6, x - 3$

4. $5a, a - 1$

5. $t^2 - 25, t + 5$

6. $x^2 - 3x - 4, x + 1$

Simplify each expression.

7. $\frac{3}{x} + \frac{5}{y}$

8. $\frac{3}{8p^2r} + \frac{5}{4p^2r}$

9. $\frac{2c - 7}{3} + 4$

10. $\frac{2}{m^2p} + \frac{5}{p}$

11. $\frac{12}{5y^2} - \frac{2}{5yz}$

12. $\frac{7}{4gh} + \frac{3}{4h^2}$

13. $\frac{2}{a + 2} - \frac{3}{2a}$

14. $\frac{5}{3b + d} - \frac{2}{3bd}$

15. $\frac{3}{w - 3} - \frac{2}{w^2 - 9}$

16. $\frac{3t}{2 - x} + \frac{5}{x - 2}$

17. $\frac{k}{k - n} - \frac{k}{n - k}$

18. $\frac{4z}{z - 4} + \frac{z + 4}{z + 1}$

19. $\frac{1}{x^2 + 2x + 1} + \frac{x}{x + 1}$

20. $\frac{2x + 1}{x - 5} - \frac{4}{x^2 - 3x - 10}$

21. $\frac{n}{n - 3} + \frac{2n + 2}{n^2 - 2n - 3}$

22. $\frac{3}{y^2 + y - 12} - \frac{2}{y^2 + 6y + 8}$

NAME ______________________ DATE ____________ PERIOD ________

9-2 Practice

Adding and Subtracting Rational Expressions

Find the LCM of each set of polynomials.

1. $x^2y,\ xy^3$

2. $a^2b^3c,\ abc^4$

3. $x + 1,\ x + 3$

4. $g - 1,\ g^2 + 3g - 4$

5. $2r + 2,\ r^2 + r,\ r + 1$

6. $3,\ 4w + 2,\ 4w^2 - 1$

7. $x^2 + 2x - 8,\ x + 4$

8. $x^2 - x - 6,\ x^2 + 6x + 8$

9. $d^2 + 6d + 9,\ 2(d^2 - 9)$

Simplify each expression.

10. $\frac{5}{6ab} - \frac{7}{8a}$

11. $\frac{5}{12x^4y} - \frac{1}{5x^2y^3}$

12. $\frac{1}{6c^2d} + \frac{3}{4cd^3}$

13. $\frac{4m}{3mn} + 2$

14. $2x - 5 - \frac{x - 8}{x + 4}$

15. $\frac{4}{a - 3} + \frac{9}{a - 5}$

16. $\frac{16}{x^2 - 16} + \frac{2}{x + 4}$

17. $\frac{2 - 5m}{m - 9} + \frac{4m - 5}{9 - m}$

18. $\frac{y - 5}{y^2 - 3y - 10} + \frac{y}{y^2 + y - 2}$

19. $\frac{5}{2x - 12} - \frac{20}{x^2 - 4x - 12}$

20. $\frac{2p - 3}{p^2 - 5p + 6} - \frac{5}{p^2 - 9}$

21. $\frac{1}{5n} - \frac{3}{4} + \frac{7}{10n}$

22. $\frac{2a}{a - 3} - \frac{2a}{a + 3} + \frac{36}{a^2 - 9}$

23. $\dfrac{\frac{2}{x - y} + \frac{1}{x + y}}{\frac{1}{x - y}}$

24. $\dfrac{\frac{r + 6}{r} - \frac{1}{r + 2}}{\frac{r^2 + 4r + 3}{r^2 + 2r}}$

25. GEOMETRY The expressions $\frac{5x}{2}$, $\frac{20}{x + 4}$, and $\frac{10}{x - 4}$ represent the lengths of the sides of a triangle. Write a simplified expression for the perimeter of the triangle.

26. KAYAKING Mai is kayaking on a river that has a current of 2 miles per hour. If r represents her rate in calm water, then $r + 2$ represents her rate with the current, and $r - 2$ represents her rate against the current. Mai kayaks 2 miles downstream and then back to her starting point. Use the formula for time, $t = \frac{d}{r}$, where d is the distance, to write a simplified expression for the total time it takes Mai to complete the trip.

NAME ______________________ DATE __________ PERIOD __________

9-3 Skills Practice

Graphing Reciprocal Functions

Identify the asymptotes, domain, and range of each function.

1.

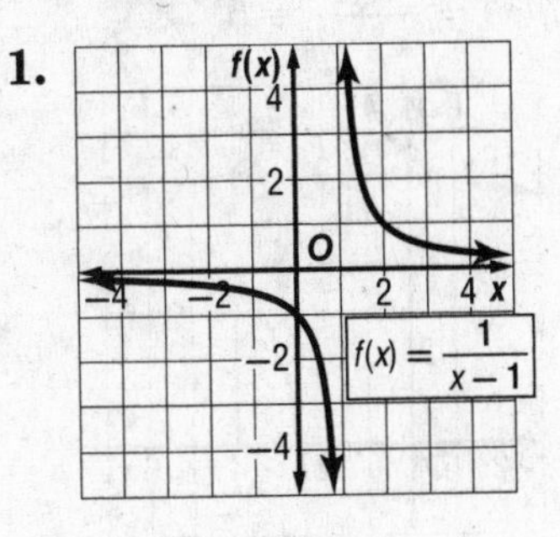

2.

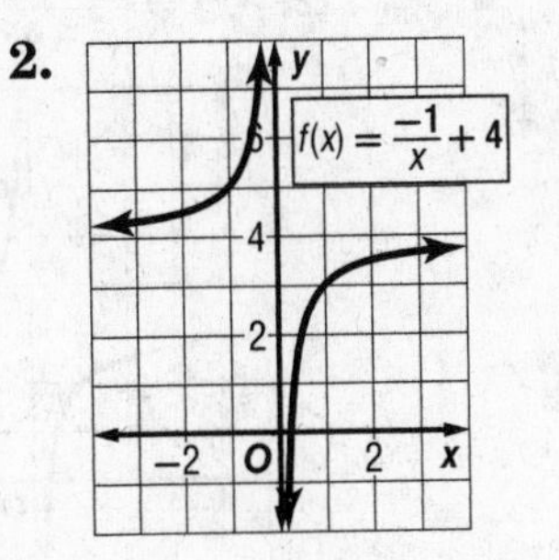

Graph each function. State the domain and range.

3. $f(x) = \frac{1}{x + 3} - 3$

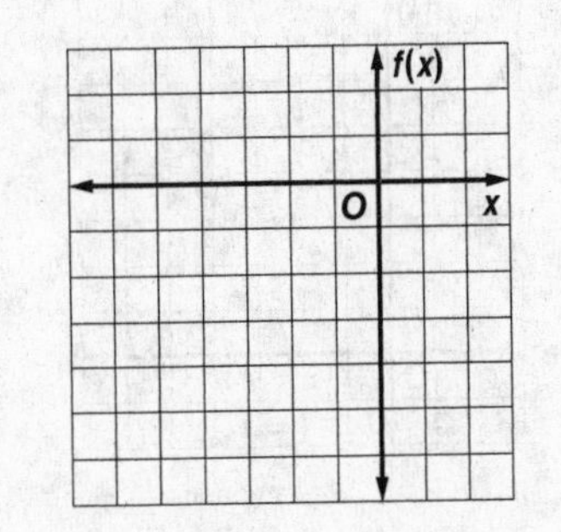

4. $f(x) = \frac{-1}{x + 5} - 6$

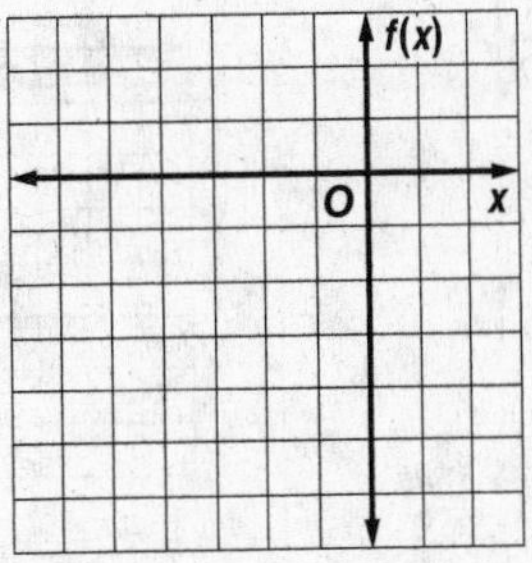

5. $f(x) = \frac{-1}{x + 1} + 3$

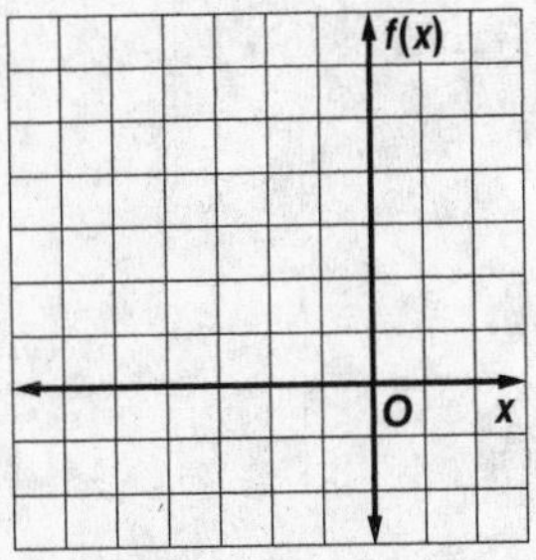

6. $f(x) = \frac{1}{x + 4} - 2$

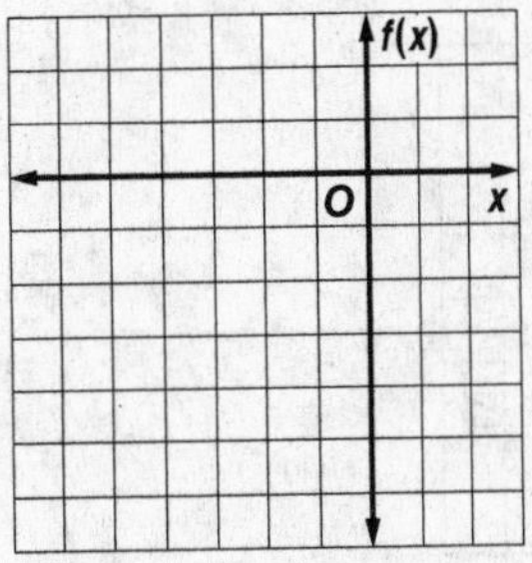

NAME ______________________ DATE ____________ PERIOD ____________

9-3 Practice

Graphing Reciprocal Functions

Identify the asymptotes, domain, and range of each function.

1. $f(x) = \frac{1}{x-1} - 3$

2. $f(x) = \frac{1}{x+1} + 3$

3. $f(x) = \frac{-3}{x-2} + 5$

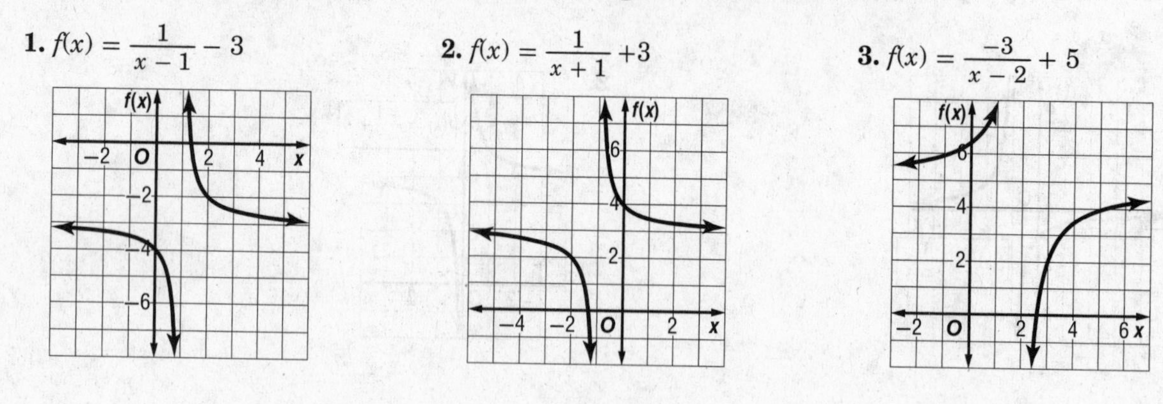

Graph each function. State the domain and range.

4. $f(x) = \frac{1}{x+1} - 5$

5. $f(x) = \frac{-1}{x-3} - 4$

6. $f(x) = \frac{3}{x-2} + 4$

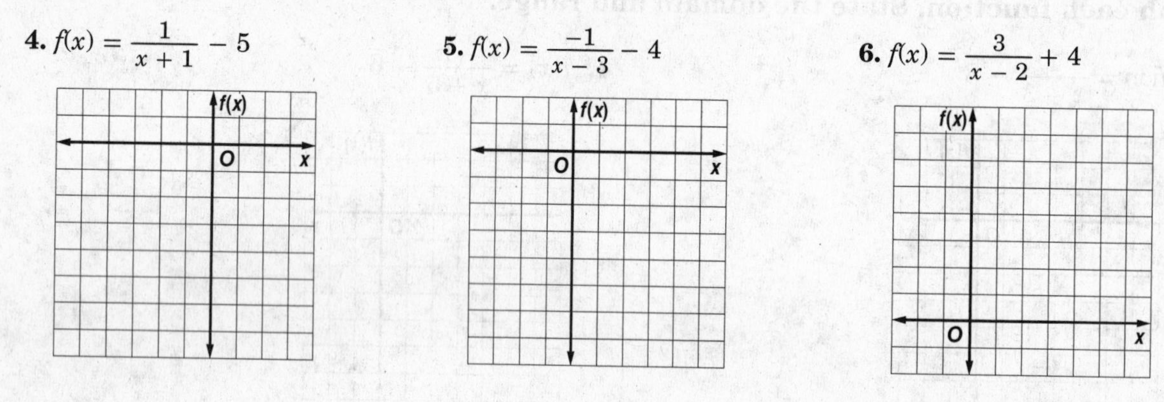

7. RACE Kate enters a 120-mile bicycle race. Her basic rate is 10 miles per hour, but Kate will average x miles per hour faster than that. Write and graph an equation relating x (Kate's speed beyond 10 miles per hour) to the time it would take to complete the race. If she wanted to finish the race in 4 hours instead of 5 hours, how much faster should she travel?

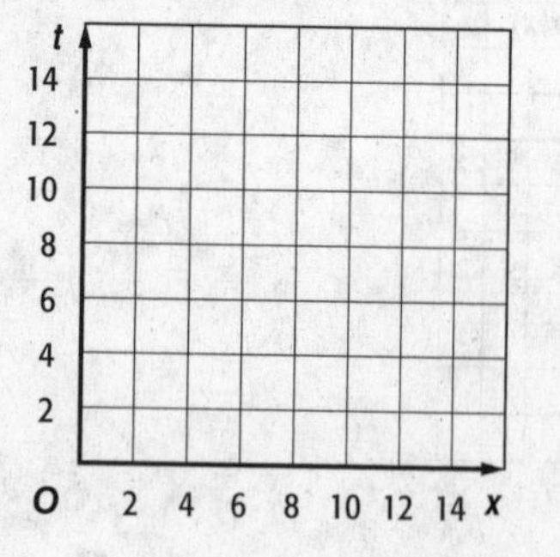

NAME ______________________ DATE ____________ PERIOD ________

9-4 Skills Practice

Graphing Rational Functions

Graph each function.

1. $f(x) = \frac{-3}{x}$

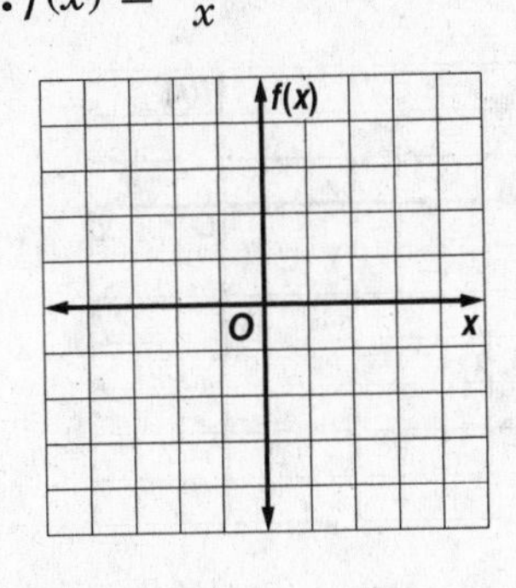

2. $f(x) = \frac{10}{x}$

3. $f(x) = \frac{-4}{x}$

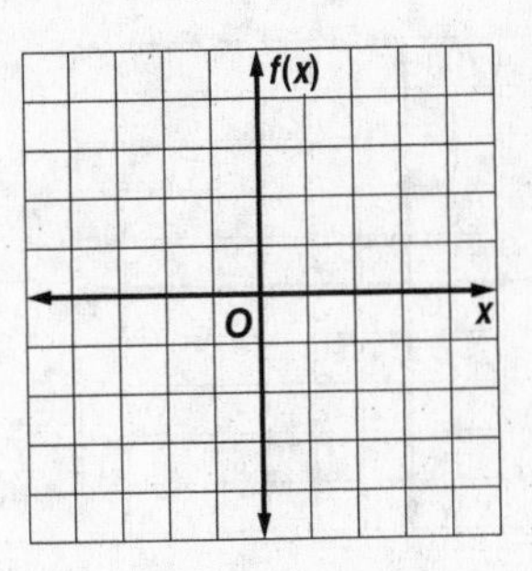

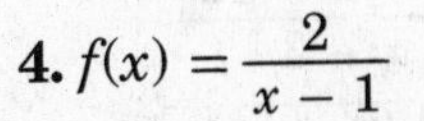

4. $f(x) = \frac{2}{x - 1}$

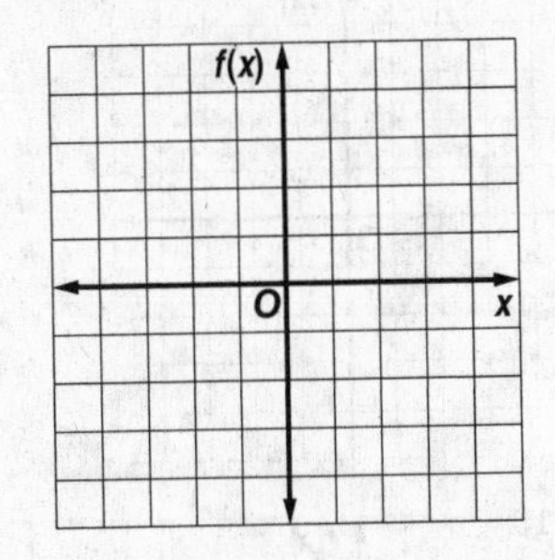

5. $f(x) = \frac{x}{x + 2}$

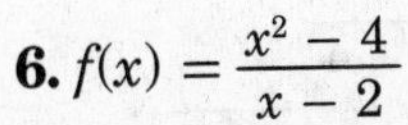

6. $f(x) = \frac{x^2 - 4}{x - 2}$

7. $f(x) = \frac{x^2 + x - 12}{x - 3}$

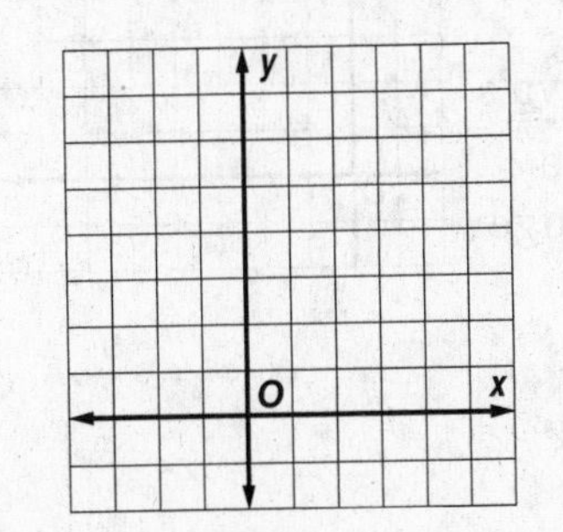

8. $f(x) = \frac{x - 1}{x^2 - 4x + 3}$

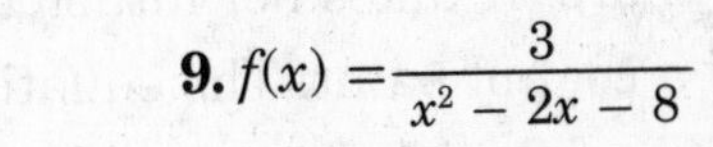

9. $f(x) = \frac{3}{x^2 - 2x - 8}$

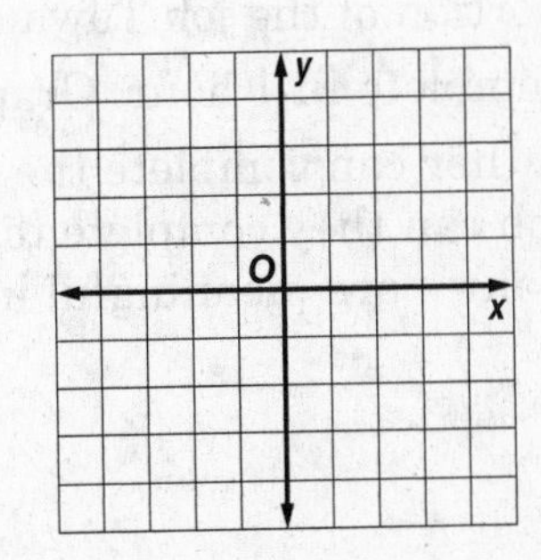

9. $f(x) = \frac{x^3}{2x + 2}$

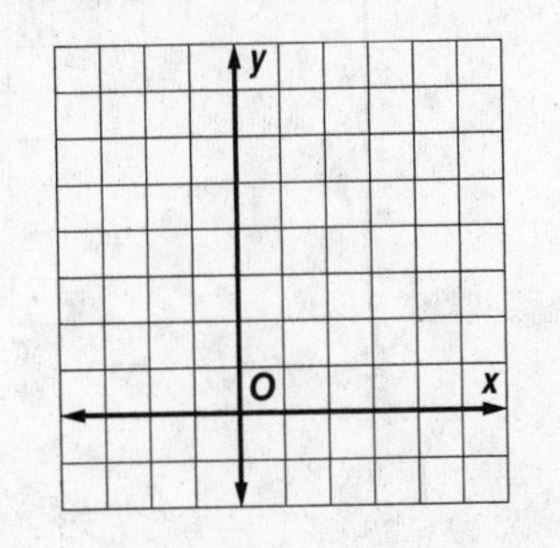

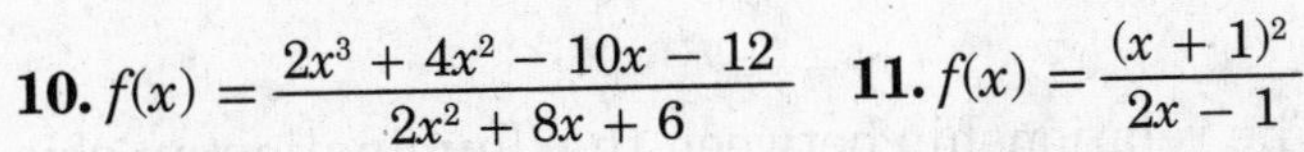

10. $f(x) = \frac{2x^3 + 4x^2 - 10x - 12}{2x^2 + 8x + 6}$

11. $f(x) = \frac{(x + 1)^2}{2x - 1}$

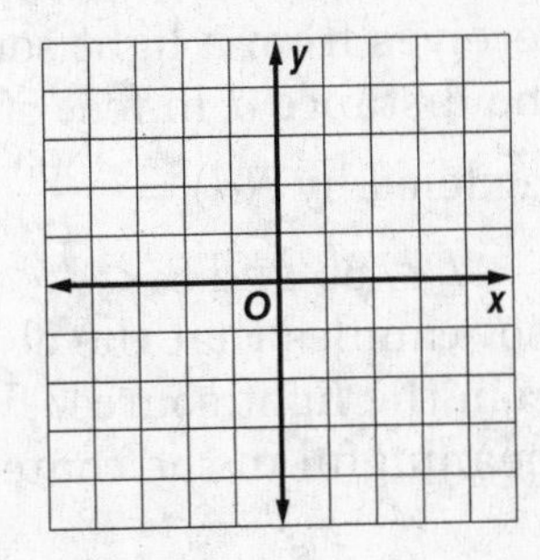

NAME ______________________________ DATE ____________ PERIOD ____________

9-4 Practice

Graphing Rational Functions

Graph each function.

1. $f(x) = \dfrac{-4}{x - 2}$

2. $f(x) = \dfrac{x - 3}{x - 2}$

3. $f(x) = \dfrac{3x}{(x + 3)^2}$

4. $f(x) = \dfrac{2x^2 + 5}{6x - 4}$

5. $f(x) = \dfrac{x^2 + 2x - 8}{x - 2}$

6. $f(x) = \dfrac{x^2 - 7x + 12}{x - 3}$

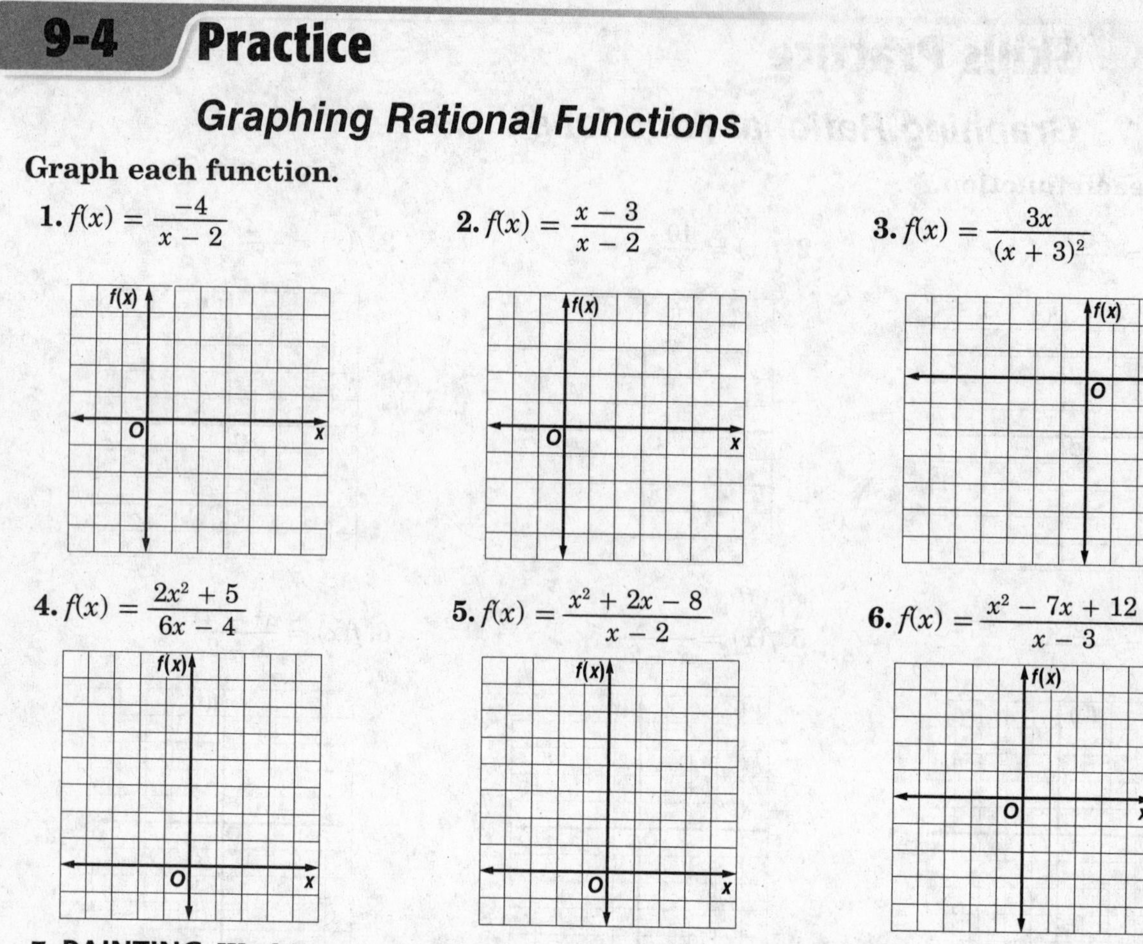

7. PAINTING Working alone, Tawa can give the shed a coat of paint in 6 hours. It takes her father x hours working alone to give the shed a coat of paint. The equation $f(x) = \dfrac{6 + x}{6x}$ describes the portion of the job Tawa and her father working together can complete in 1 hour. Graph $f(x) = \dfrac{6 + x}{6x}$ for $x > 0$, $f(x) > 0$. If Tawa's father can complete the job in 4 hours alone, what portion of the job can they complete together in 1 hour? What domain and range values are meaningful in the context of the problem?

8. LIGHT The relationship between the illumination an object receives from a light source of I foot-candles and the square of the distance d in feet of the object from the source can be modeled by $I(d) = \dfrac{4500}{d^2}$. Graph the function $I(d) = \dfrac{4500}{d^2}$ for $0 < I \leq 80$ and $0 < d \leq 80$. What is the illumination in foot-candles that the object receives at a distance of 20 feet from the light source? What domain and range values are meaningful in the context of the problem?

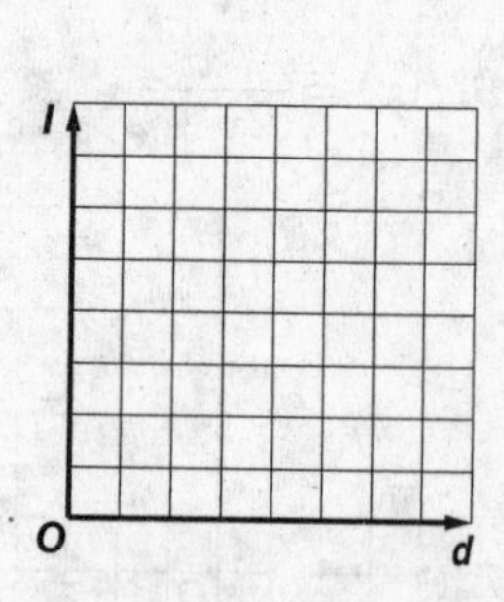

NAME ______________________ DATE ____________ PERIOD ____________

9-5 Skills Practice

Variation Functions

State whether each equation represents a *direct, joint, inverse, or combined* variation. Then name the constant of variation.

1. $c = 12m$

2. $p = \frac{4}{q}$

3. $A = \frac{1}{2}bh$

4. $rw = 15$

5. $y = 2rgt$

6. $f = 5280m$

7. $y = 0.2d$

8. $vz = -25$

9. $t = 16rh$

10. $R = \frac{8}{w}$

11. $b = \frac{1}{3}a$

12. $C = 2\pi r$

13. If y varies directly as x and $y = 35$ when $x = 7$, find y when $x = 11$.

14. If y varies directly as x and $y = 360$ when $x = 180$, find y when $x = 270$.

15. If y varies directly as x and $y = 540$ when $x = 10$, find x when $y = 1080$.

16. If y varies directly as x and $y = 12$ when $x = 72$, find x when $y = 9$.

17. If y varies jointly as x and z and $y = 18$ when $x = 2$ and $z = 3$, find y when x is 5 and z is 6.

18. If y varies jointly as x and z and $y = -16$ when $x = 4$ and $z = 2$, find y when x is -1 and z is 7.

19. If y varies jointly as x and z and $y = 120$ when $x = 4$ and $z = 6$, find y when x is 3 and z is 2.

20. If y varies inversely as x and $y = 2$ when $x = 2$, find y when $x = 1$.

21. If y varies inversely as x and $y = 6$ when $x = 5$, find y when $x = 10$.

22. If y varies inversely as x and $y = 3$ when $x = 14$, find x when $y = 6$.

23. If y varies directly as z and inversely as x and $y = 27$ and $z = -3$ when $x = 2$, find x when $y = 9$ and $z = 5$.

24. If y varies directly as z and inversely as x and $y = -15$ and $z = 5$ when $x = 5$, find x when $y = -36$ and $z = -3$.

NAME ______________________ DATE ____________ PERIOD ____________

9-5 Practice

Variation Functions

State whether each equation represents a *direct, joint, inverse,* or *combined* variation. Then name the constant of variation.

1. $u = 8wz$

2. $p = 4s$

3. $L = \frac{5}{k}$

4. $xy = 4.5$

5. $\frac{C}{d} = \pi$

6. $2d = mn$

7. $\frac{1.25}{g} = h$

8. $y = \frac{3}{4x}$

9. If y varies directly as x and $y = 8$ when $x = 2$, find y when $x = 6$.

10. If y varies directly as x and $y = -16$ when $x = 6$, find x when $y = -4$.

11. If y varies directly as x and $y = 132$ when $x = 11$, find y when $x = 33$.

12. If y varies directly as x and $y = 7$ when $x = 1.5$, find y when $x = 4$.

13. If y varies jointly as x and z and $y = 24$ when $x = 2$ and $z = 1$, find y when x is 12 and z is 2.

14. If y varies jointly as x and z and $y = 60$ when $x = 3$ and $z = 4$, find y when x is 6 and z is 8.

15. If y varies jointly as x and z and $y = 12$ when $x = -2$ and $z = 3$, find y when x is 4 and z is -1.

16. If y varies inversely as x and $y = 16$ when $x = 4$, find y when $x = 3$.

17. If y varies inversely as x and $y = 3$ when $x = 5$, find x when $y = 2.5$.

18. If y varies directly as z and inversely as x and $y = -18$ and $z = 3$ when $x = 6$, find y when $x = 5$ and $z = -5$.

19. If y varies directly as z and inversely as x and $y = 10$ and $z = 5$ when $x = 12.5$, find z when $y = 37.5$ and $x = 2$.

20. GASES The volume V of a gas varies inversely as its pressure P. If $V = 80$ cubic centimeters when $P = 2000$ millimeters of mercury, find V when $P = 320$ millimeters of mercury.

21. SPRINGS The length S that a spring will stretch varies directly with the weight F that is attached to the spring. If a spring stretches 20 inches with 25 pounds attached, how far will it stretch with 15 pounds attached?

22. GEOMETRY The area A of a trapezoid varies jointly as its height and the sum of its bases. If the area is 480 square meters when the height is 20 meters and the bases are 28 meters and 20 meters, what is the area of a trapezoid when its height is 8 meters and its bases are 10 meters and 15 meters?

NAME ______________________ DATE ____________ PERIOD ____________

9-6 Skills Practice

Solving Rational Equations and Inequalities

Solve each equation. Check your solution.

1. $\frac{x}{x-1} = \frac{1}{2}$

2. $2 = \frac{4}{n} + \frac{1}{3}$

3. $\frac{9}{3x} = \frac{-6}{2}$

4. $3 - z = \frac{2}{z}$

5. $\frac{2}{d+1} = \frac{1}{d-2}$

6. $\frac{r-3}{5} = \frac{8}{r}$

7. $\frac{2x+3}{x+1} = \frac{3}{2}$

8. $\frac{-12}{y} = y - 7$

9. $\frac{15}{x} + \frac{9x-7}{x+2} = 9$

10. $\frac{3b-2}{b+1} = 4 - \frac{b+2}{b-1}$

11. $2 = \frac{5}{2q} + \frac{2q}{q+1}$

12. $8 - \frac{4}{z} = \frac{8z-8}{z+2}$

13. $\frac{1}{n+3} + \frac{5}{n^2-9} = \frac{2}{n-3}$

14. $\frac{1}{w+2} + \frac{1}{w-2} = \frac{4}{w^2-4}$

15. $\frac{x-8}{2x+2} + \frac{x}{2x+2} = \frac{2x-3}{x+1}$

16. $\frac{12p+19}{p^2+7p+12} - \frac{3}{p+3} = \frac{5}{p+4}$

17. $\frac{2f}{f^2-4} + \frac{1}{f-2} = \frac{2}{f+2}$

18. $\frac{8}{t^2-9} + \frac{4}{t+3} = \frac{2}{t-3}$

Solve each inequality. Check your solutions.

19. $\frac{x-2}{x+4} > \frac{x+1}{x+10}$

20. $\frac{3}{k} - \frac{4}{3k} > 0$

21. $2 - \frac{3}{v} < \frac{5}{v}$

22. $n + \frac{3}{n} < \frac{12}{n}$

23. $\frac{1}{2m} - \frac{3}{m} < -\frac{5}{2}$

24. $\frac{1}{2x} < \frac{2}{x} - 1$

NAME ______________________ DATE ____________ PERIOD ____________

9-6 Practice

Solving Rational Equations and Inequalities

Solve each equation or inequality. Check your solutions.

1. $\frac{12}{x} + \frac{3}{4} = \frac{3}{2}$

2. $\frac{x}{x-1} - 1 = \frac{x}{2}$

3. $\frac{p+10}{p^2-2} = \frac{4}{p}$

4. $\frac{s}{s+2} + s = \frac{5s+8}{s+2}$

5. $\frac{5}{y-5} = \frac{y}{y-5} - 1$

6. $\frac{1}{3x-2} + \frac{5}{x} = 0$

7. $\frac{5}{t} < \frac{9}{2t+1}$

8. $\frac{1}{2h} + \frac{5}{h} = \frac{3}{h-1}$

9. $\frac{4}{w-2} = \frac{-1}{w+3}$

10. $5 - \frac{3}{a} < \frac{7}{a}$

11. $\frac{4}{5x} + \frac{1}{10} < \frac{3}{2x}$

12. $8 + \frac{3}{y} > \frac{19}{y}$

13. $\frac{4}{p} + \frac{1}{3p} < \frac{1}{5}$

14. $\frac{6}{x-1} = \frac{4}{x-2} + \frac{2}{x+1}$

15. $g + \frac{g}{g-2} = \frac{2}{g-2}$

16. $b + \frac{2b}{b-1} = 1 - \frac{b-3}{b-1}$

17. $\frac{1}{n+2} + \frac{1}{n-2} = \frac{3}{n^2-4}$

18. $\frac{c+1}{c-3} = 4 - \frac{12}{c^2-2c-3}$

19. $\frac{3}{k-3} + \frac{4}{k-4} = \frac{25}{k^2-7k+12}$

20. $\frac{4v}{v-1} - \frac{5v}{v-2} = \frac{2}{v^2-3v+2}$

21. $\frac{y}{y+2} + \frac{7}{y-5} = \frac{14}{y^2-3y-10}$

22. $\frac{x^2+4}{x^2-4} + \frac{x}{2-x} = \frac{2}{x+2}$

23. $\frac{r}{r+4} + \frac{4}{r-4} = \frac{r^2+16}{r^2-16}$

24. $3 = \frac{6a-1}{2a+7} + \frac{22}{a+5}$

27. BASKETBALL Kiana has made 9 of 19 free throws so far this season. Her goal is to make 60% of her free throws. If Kiana makes her next x free throws in a row, the function $f(x) = \frac{9+x}{19+x}$ represents Kiana's new ratio of free throws made. How many successful free throws in a row will raise Kiana's percent made to 60%? Is this a reasonable answer? Explain.

28. OPTICS The lens equation $\frac{1}{p} + \frac{1}{q} = \frac{1}{f}$ relates the distance p of an object from a lens, the distance q of the image of the object from the lens, and the focal length f of the lens. What is the distance of an object from a lens if the image of the object is 5 centimeters from the lens and the focal length of the lens is 4 centimeters? Is this a reasonable answer? Explain.

NAME ______________________ DATE ____________ PERIOD ____________

10-1 Skills Practice

Midpoint and Distance Formulas

Find the midpoint of the line segment with endpoints at the given coordinates.

1. (4, −1), (−4, 1)

2. (−1, 4), (5, 2)

3. (3, 4), (5, 4)

4. (6, 2), (2, −1)

5. (3, 9), (−2, −3)

6. (−3, 5), (−3, −8)

7. (3, 2), (−5, 0)

8. (3, −4), (5, 2)

9. (−5, −9), (5, 4)

10. (−11, 14), (0, 4)

11. (3, −6), (−8, −3)

12. (0, 10), (−2, −5)

Find the distance between each pair of points with the given coordinates.

13. (4, 12), (−1, 0)

14. (7, 7), (−5, −2)

15. (−1, 4), (1, 4)

16. (11, 11), (8, 15)

17. (1, −6), (7, 2)

18. (3, −5), (3, 4)

19. (2, 3), (3, 5)

20. (−4, 3), (−1, 7)

21. (−5, −5), (3, 10)

22. (3, 9), (−2, −3)

23. (6, −2), (−1, 3)

24. (−4, 1), (2, −4)

25. (0, −3), (4, 1)

26. (−5, −6), (2, 0)

NAME ______________________ DATE ____________ PERIOD ________

10-1 Practice

Midpoint and Distance Formulas

Find the midpoint of the line segment with endpoints at the given coordinates.

1. (8, −3), (−6, −11)

2. (−14, 5), (10, 6)

3. (−7, −6), (1, −2)

4. (8, −2), (8, −8)

5. (9, −4), (1, −1)

6. (3, 3), (4, 9)

7. (4, −2), (3, −7)

8. (6, 7), (4, 4)

9. (−4, −2), (−8, 2)

10. (5, −2), (3, 7)

11. (−6, 3), (−5, −7)

12. (−9, −8), (8, 3)

13. (2.6, −4.7), (8.4, 2.5)

14. $\left(-\frac{1}{3}, 6\right), \left(\frac{2}{3}, 4\right)$

15. (−2.5, −4.2), (8.1, 4.2)

16. $\left(\frac{1}{8}, \frac{1}{2}\right), \left(-\frac{5}{8}, -\frac{1}{2}\right)$

Find the distance between each pair of points with the given coordinates.

17. (5, 2), (2, −2)

18. (−2, −4), (4, 4)

19. (−3, 8), (−1, −5)

20. (0, 1), (9, −6)

21. (−5, 6), (−6, 6)

22. (−3, 5), (12, −3)

23. (−2, −3), (9, 3)

24. (−9, −8), (−7, 8)

25. (9, 3), (9, −2)

26. (−1, −7), (0, 6)

27. (10, −3), (−2, −8)

28. (−0.5, −6), (1.5, 0)

29. $\left(\frac{2}{5}, \frac{3}{5}\right), \left(1, \frac{7}{5}\right)$

30. $(-4\sqrt{2}, -\sqrt{5}), (-5\sqrt{2}, 4\sqrt{5})$

31. GEOMETRY Circle O has a diameter $\overline{AB}$. If A is at (−6, −2) and B is at (−3, 4), find the center of the circle and the length of its diameter.

32. GEOMETRY Find the perimeter of a triangle with vertices at (1, −3), (−4, 9), and (−2, 1).

NAME ______________________ DATE ____________ PERIOD ____________

10-2 Skills Practice

Parabolas

Write each equation in standard form. Identify the vertex, axis of symmetry, and direction of opening of the parabola.

1. $y = x^2 + 2x + 2$

2. $y = x^2 - 2x + 4$

3. $y = x^2 + 4x + 1$

4. $y = -2x^2 + 12x - 14$

5. $x = 3y^2 + 6y - 5$

6. $x + y^2 - 8y = -20$

Graph each equation.

4. $y = (x - 2)^2$

5. $x = (y - 2)^2 + 3$

6. $y = -(x + 3)^2 + 4$

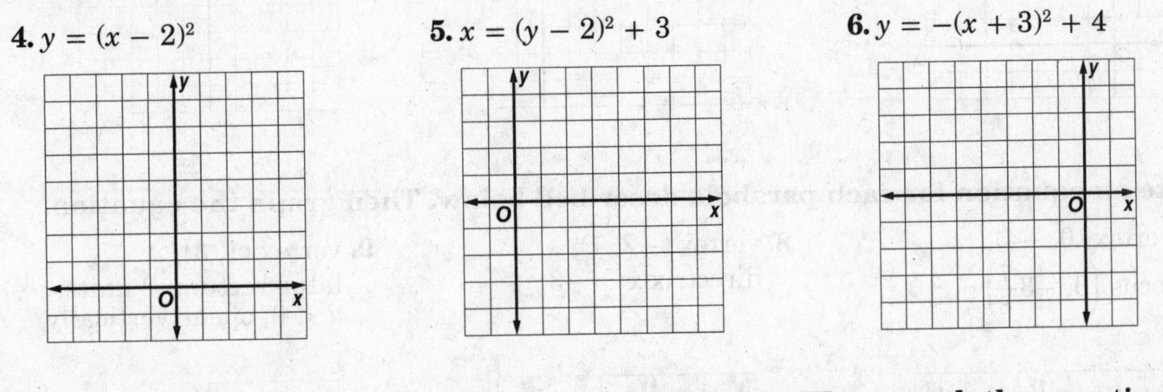

Write an equation for each parabola described below. Then graph the equation.

7. vertex (0, 0), focus $\left(0, -\frac{1}{12}\right)$

8. vertex (5, 1), focus $\left(5, \frac{5}{4}\right)$

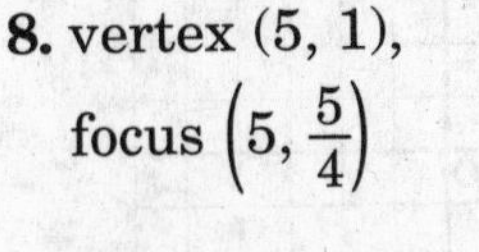

9. vertex (1, 3), directrix $x = \frac{7}{8}$

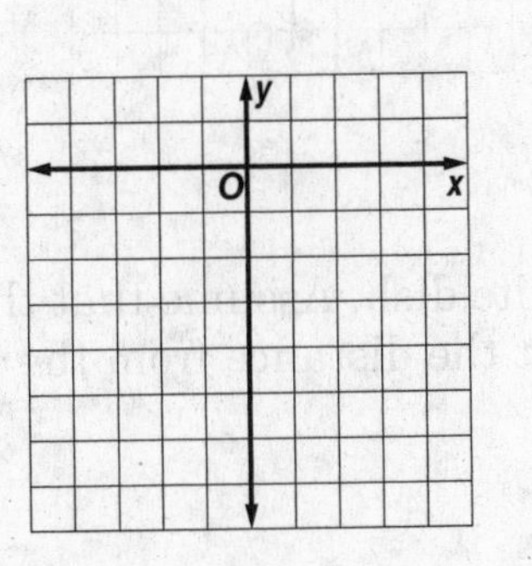

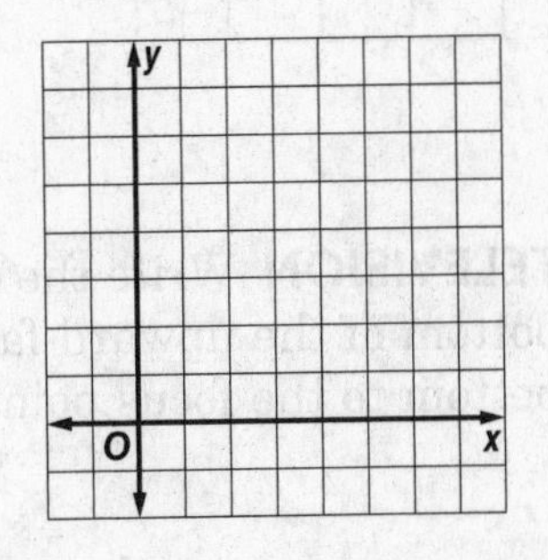

NAME ______________________________ DATE ____________ PERIOD ____________

10-2 Practice

Parabolas

Write each equation in standard form. Identify the vertex, axis of symmetry, and direction of opening of the parabola.

1. $y = 2x^2 - 12x + 19$

2. $y = \frac{1}{2}x^2 + 3x + \frac{1}{2}$

3. $y = -3x^2 - 12x - 7$

Graph each equation.

4. $y = (x - 4)^2 + 3$

5. $x = -\frac{1}{3}y^2 + 1$

6. $x = 3(y + 1)^2 - 3$

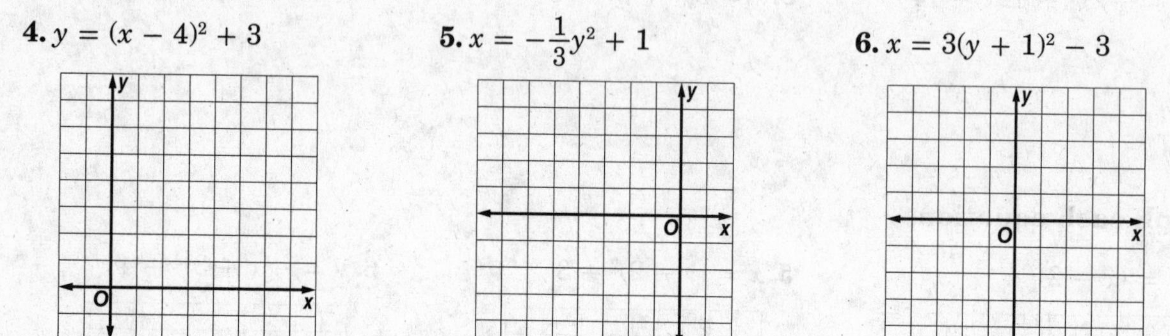

Write an equation for each parabola described below. Then graph the equation.

7. vertex $(0, -4)$, focus $\left(0, -3\frac{7}{8}\right)$

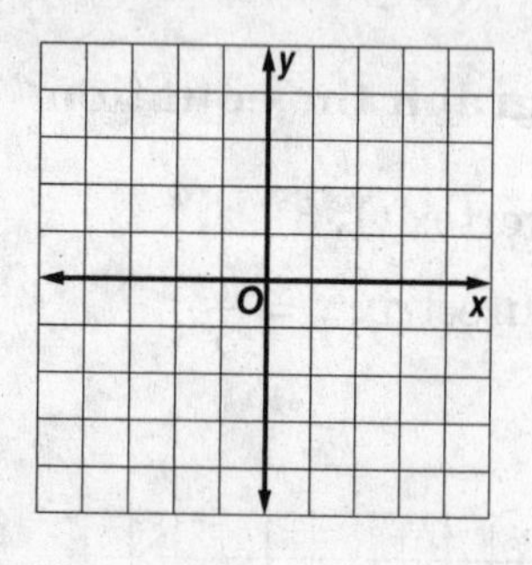

8. vertex $(-2, 1)$, directrix $x = -3$

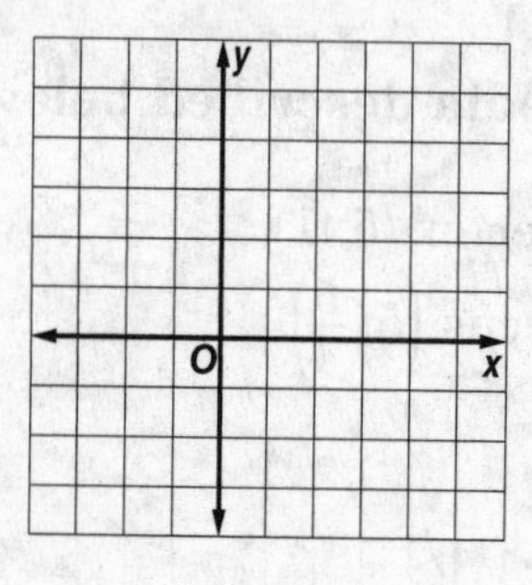

9. vertex $(1, 3)$, latus rectum: 2 units, $a < 0$, opens vertically

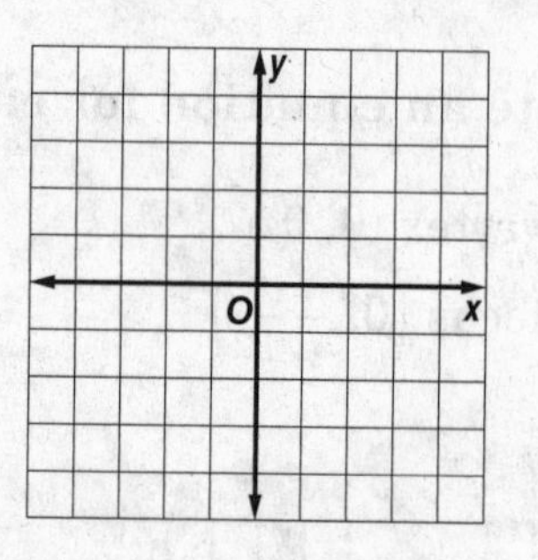

10. TELEVISION Write the equation in the form $y = ax^2$ for a satellite dish. Assume that the bottom of the upward-facing dish passes through $(0, 0)$ and that the distance from the bottom to the focus point is 8 inches.

NAME ______________________ DATE ____________ PERIOD ________

10-3 Skills Practice

Circles

Write an equation for the circle that satisfies each set of conditions.

1. center: (0, 5), $r = 1$ unit

2. center: (5, 12), $r = 8$ units

3. center: (4, 0), $r = 2$ units

4. center: (2, 2), $r = 3$ units

5. center: (4, −4), $r = 4$ units

6. center: (−6, 4), $r = 5$ units

7. endpoints of a diameter at (−12, 0) and (12, 0)

8. endpoints of a diameter at (−4, 0) and (−4, −6)

9. center at (7, −3), passes through the origin

10. center at (−4, 4), passes through (−4, 1)

11. center at (−6, −5), tangent to y-axis

12. center at (5, 1), tangent to x-axis

Find the center and radius of each circle. Then graph the circle.

13. $x^2 + y^2 = 9$

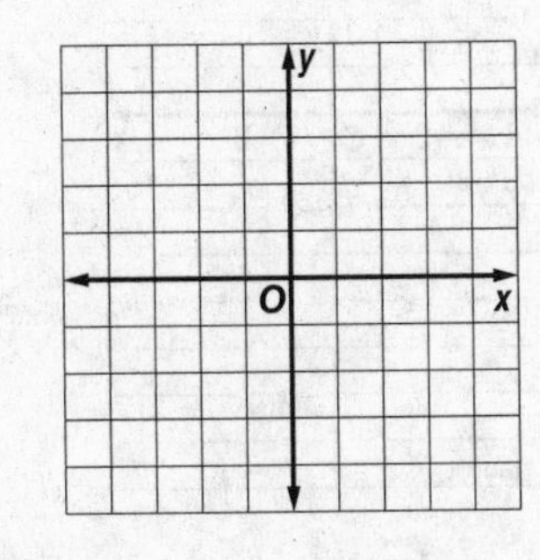

14. $(x - 1)^2 + (y - 2)^2 = 4$

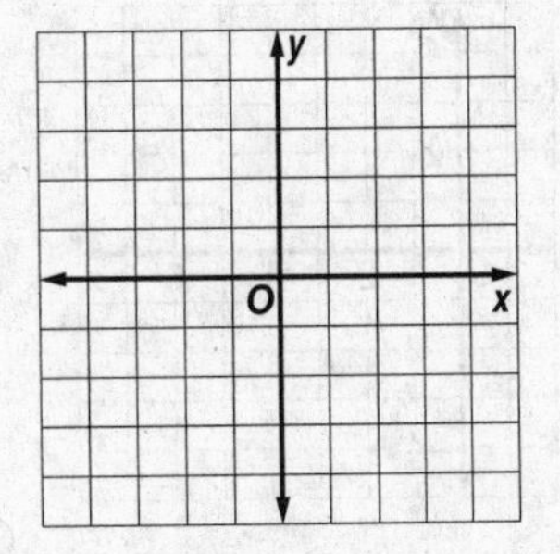

15. $(x + 1)^2 + y^2 = 16$

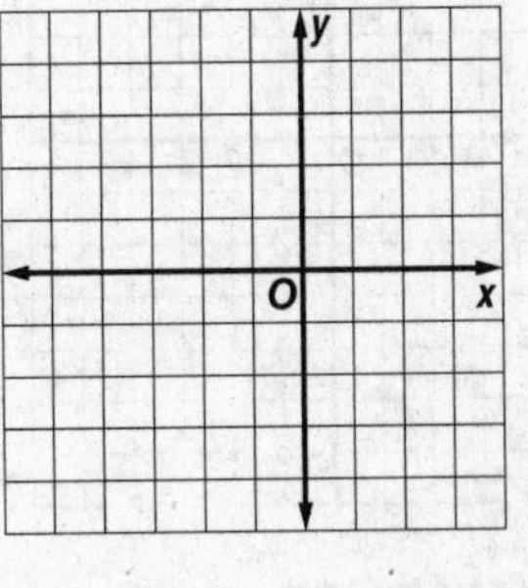

16. $x^2 + (y + 3)^2 = 81$

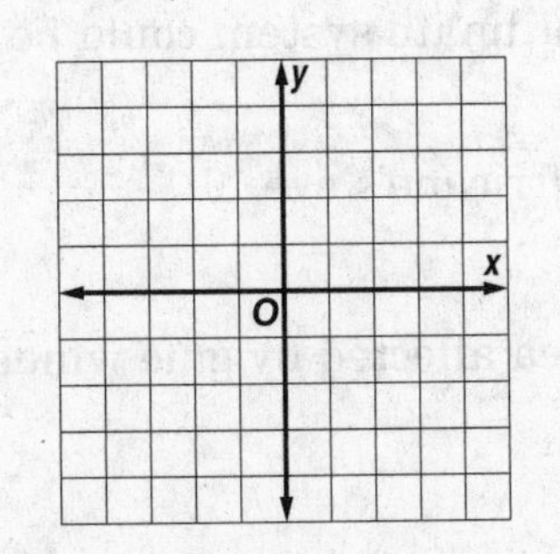

17. $(x - 5)^2 + (y + 8)^2 = 49$

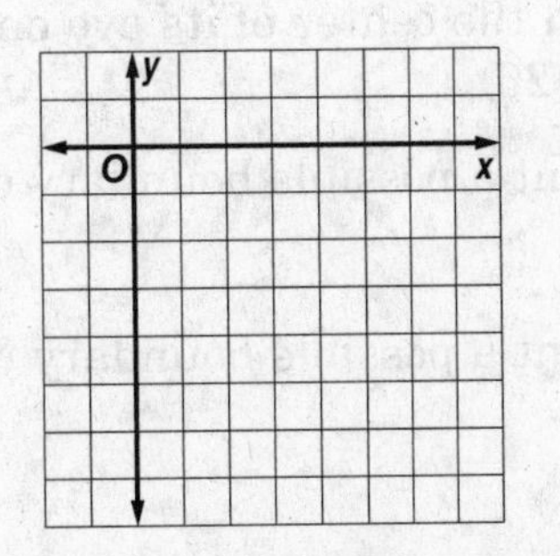

18. $x^2 + y^2 - 4y - 32 = 0$

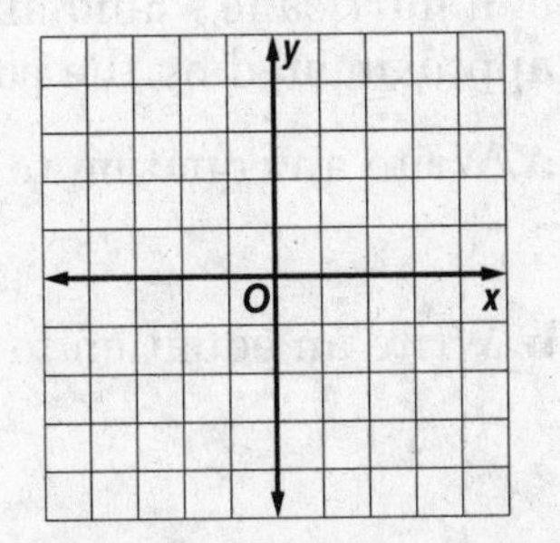

NAME ______________________________ DATE ______________ PERIOD __________

10-3 Practice

Circles

Write an equation for the circle that satisfies each set of conditions.

1. center (−4, 2), radius 8 units

2. center (0, 0), radius 4 units

3. center $\left(-\frac{1}{4}, -\sqrt{3}\right)$, radius $5\sqrt{2}$ units

4. center (2.5, 4.2), radius 0.9 units

5. endpoints of a diameter at (−2, −9) and (0, −5)

6. center at (−9, −12), passes through (−4, −5)

7. center at (−6, 5), tangent to x-axis

Find the center and radius of each circle. Then graph the circle.

8. $(x + 3)^2 + y^2 = 16$

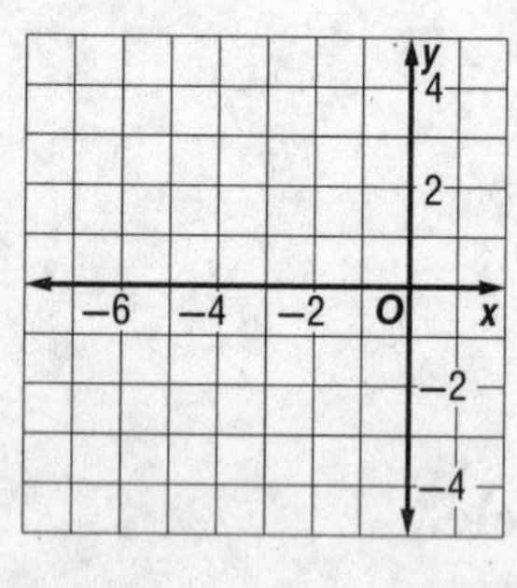

9. $3x^2 + 3y^2 = 12$

10. $x^2 + y^2 + 2x + 6y = 26$

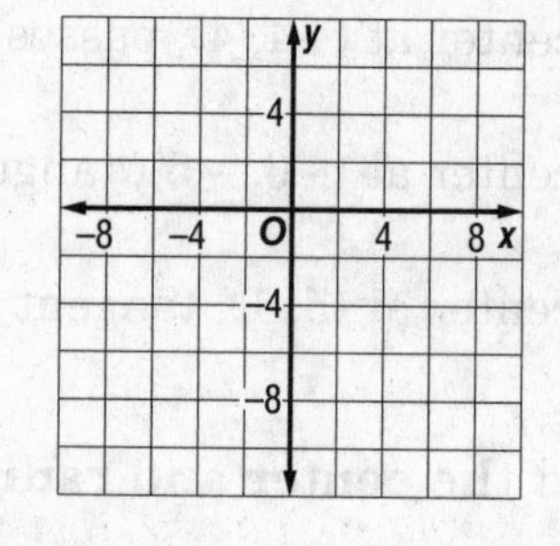

11. $(x - 1)^2 + y^2 + 4y = 12$

12. $x^2 - 6x + y^2 = 0$

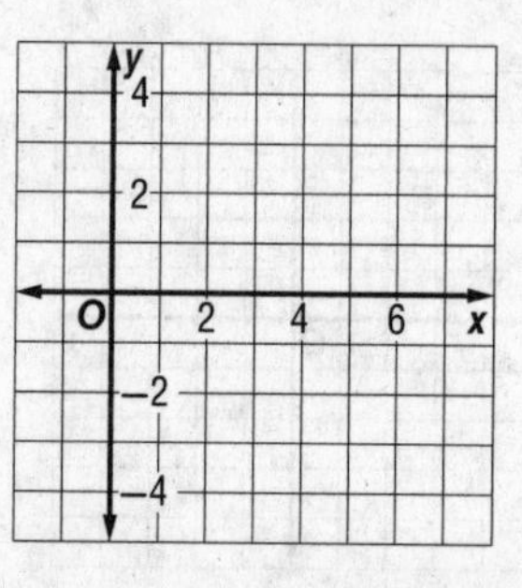

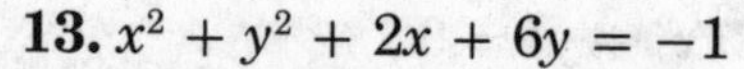

13. $x^2 + y^2 + 2x + 6y = -1$

14. WEATHER On average, the circular eye of a hurricane is about 15 miles in diameter. Gale winds can affect an area up to 300 miles from the storm's center. A satellite photo of a hurricane's landfall showed the center of its eye on one coordinate system could be approximated by the point (80, 26).

a. Write an equation to represent a possible boundary of the hurricane's eye.

b. Write an equation to represent a possible boundary of the area affected by gale winds.

10-4 Skills Practice

Ellipses

Write an equation for each ellipse.

1.

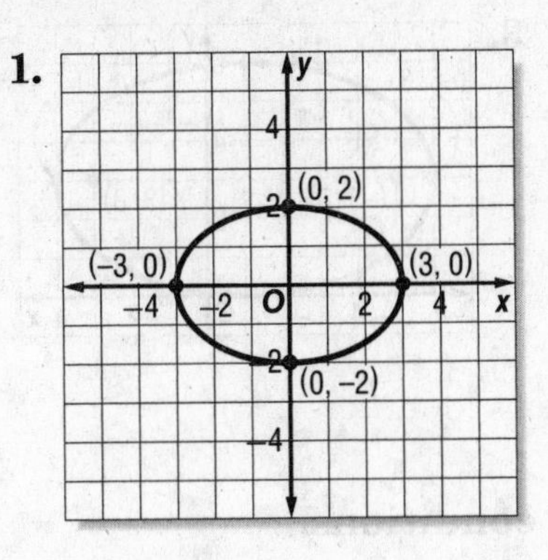

2.

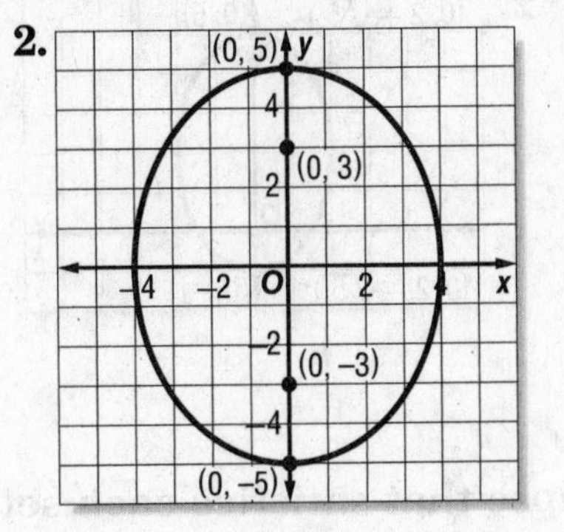

3.

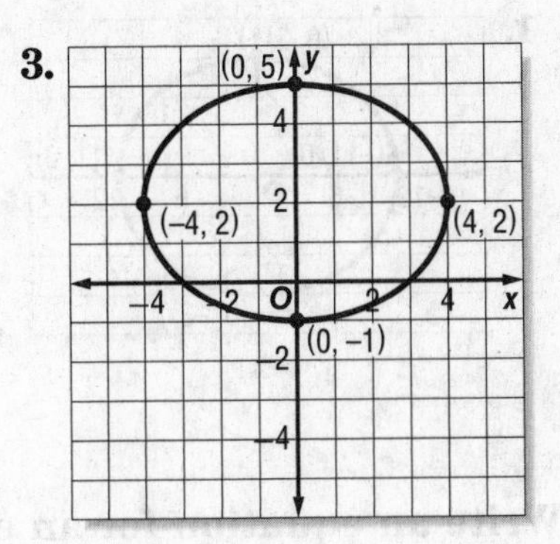

Write an equation for an ellipse that satisfies each set of conditions.

4. endpoints of major axis at (0, 6) and (0, –6), endpoints of minor axis at (–3, 0) and (3, 0)

5. endpoints of major axis at (2, 6) and (8, 6), endpoints of minor axis at (5, 4) and (5, 8)

6. endpoints of major axis at (7, 3) and (7, 9), endpoints of minor axis at (5, 6) and (9, 6)

7. major axis 12 units long and horizontal, minor axis 4 units long, center at (0, 0)

8. endpoints of major axis at (–6, 0) and (6, 0), foci at $(-\sqrt{32}, 0)$ and $(\sqrt{32}, 0)$

9. endpoints of major axis at (0, 12) and (0, –12), foci at $(0, \sqrt{23})$ and $(0, -\sqrt{23})$

Find the coordinates of the center and foci and the lengths of the major and minor axes for the ellipse with the given equation. Then graph the ellipse.

10. $\frac{y^2}{100} + \frac{x^2}{81} = 1$

11. 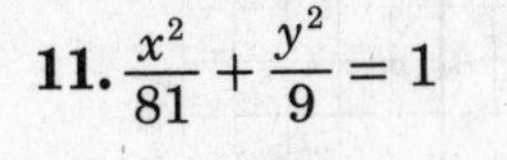
$\frac{x^2}{81} + \frac{y^2}{9} = 1$

12. $\frac{y^2}{49} + \frac{x^2}{25} = 1$

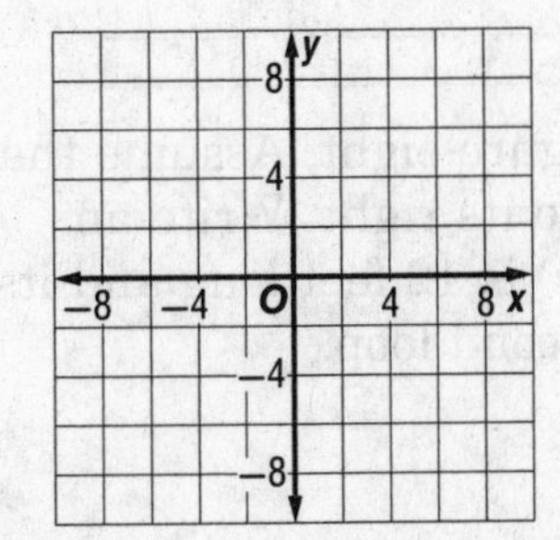

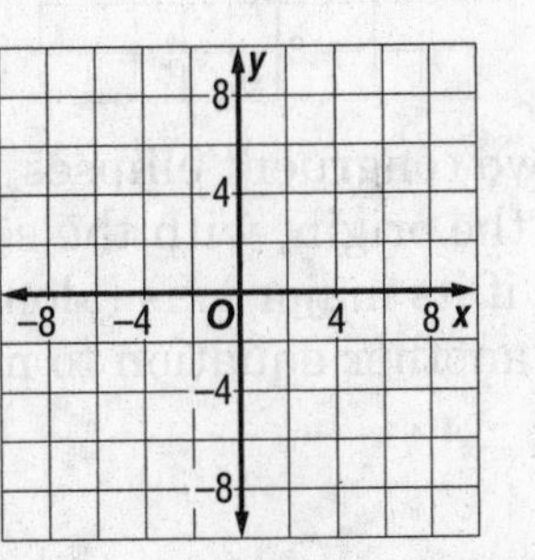

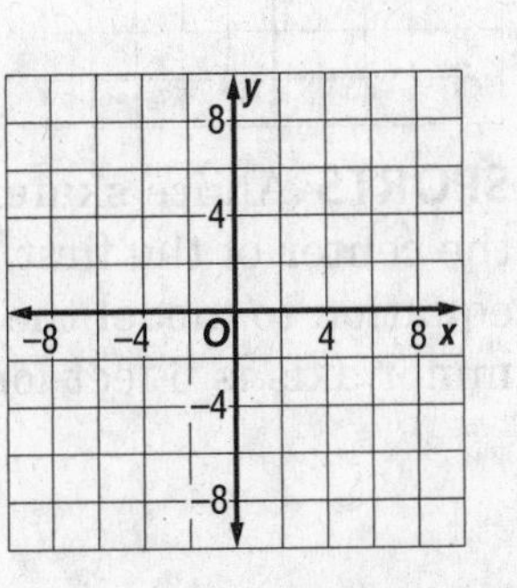

NAME ______________________ DATE __________ PERIOD __________

10-4 Practice

Ellipses

Write an equation for each ellipse.

1.

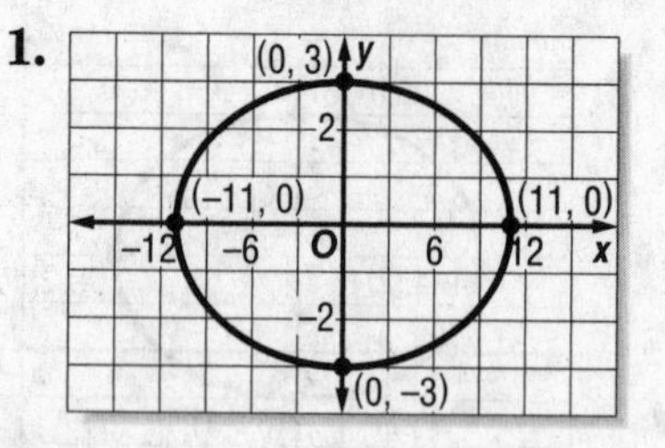

2.

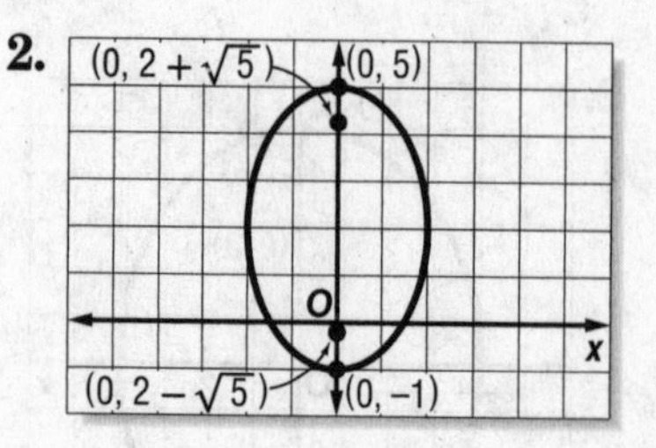

3.

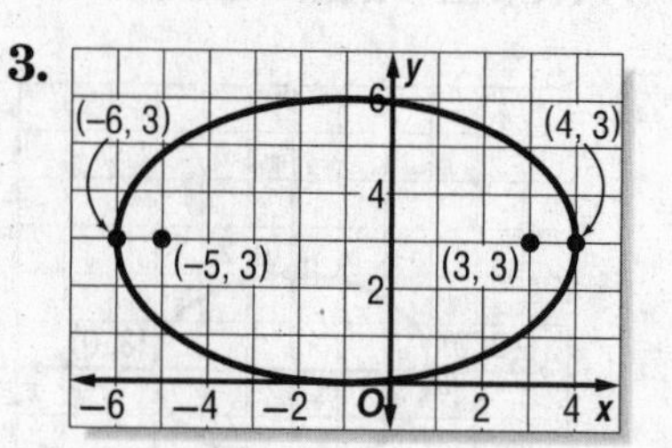

Write an equation for an ellipse that satisfies each set of conditions.

4. endpoints of major axis at $(-9, 0)$ and $(9, 0)$, endpoints of minor axis at $(0, 3)$ and $(0, -3)$

5. endpoints of major axis at $(4, 2)$ and $(4, -8)$, endpoints of minor axis at $(1, -3)$ and $(7, -3)$

6. major axis 20 units long and parallel to x-axis, minor axis 10 units long, center at $(2, 1)$

7. major axis 10 units long, minor axis 6 units long and parallel to x-axis, center at $(2, -4)$

8. major axis 16 units long, center at $(0, 0)$, foci at $(0, 2\sqrt{15})$ and $(0, -2\sqrt{15})$

9. endpoints of minor axis at $(0, 2)$ and $(0, -2)$, foci at $(-4, 0)$ and $(4, 0)$

Find the coordinates of the center and foci and the lengths of the major and minor axes for the ellipse with the given equation. Then graph the ellipse.

10. $\frac{y^2}{16} + \frac{x^2}{9} = 1$

11. $\frac{(y-1)^2}{36} + \frac{(x-3)^2}{1} = 1$

12. $\frac{(x+4)^2}{49} + \frac{(y+3)^2}{25} = 1$

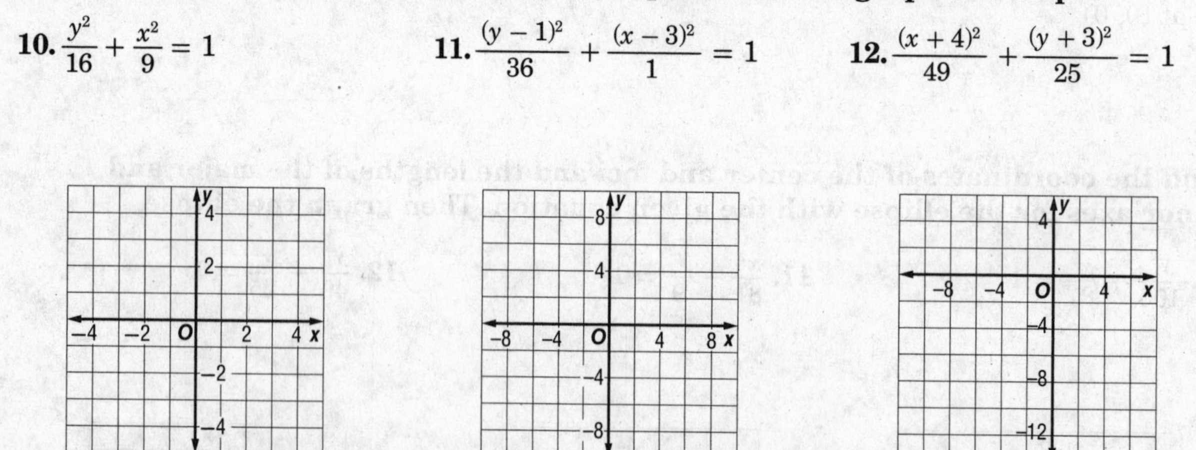

13. SPORTS An ice skater traces two congruent ellipses to form a figure eight. Assume that the center of the first loop is at the origin, with the second loop to its right. Write an equation to model the first loop if its major axis (along the x-axis) is 12 feet long and its minor axis is 6 feet long. Write another equation to model the second loop.

NAME ______________________ DATE ____________ PERIOD ____________

10-5 Skills Practice

Hyperbolas

Write an equation for each hyperbola.

1.

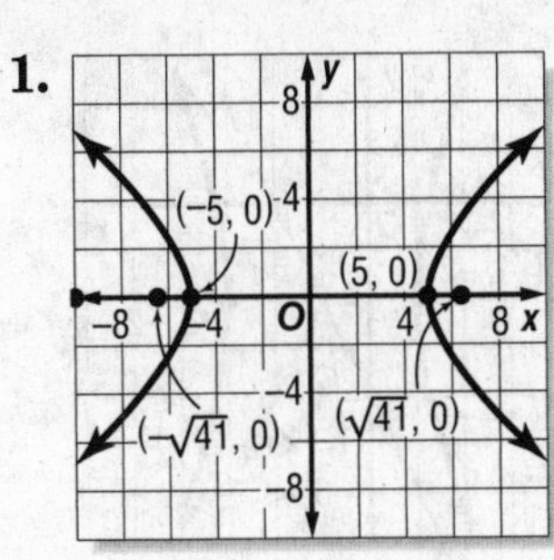

2.

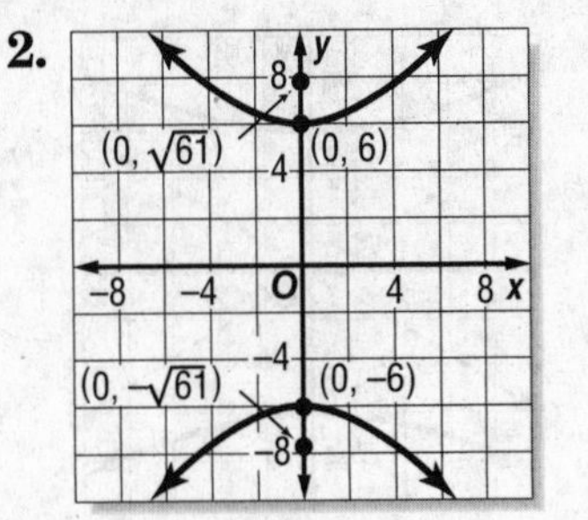

3.

Write an equation for the hyperbola that satisfies each set of conditions.

4. vertices (−4, 0) and (4, 0), conjugate axis of length 8

5. vertices (0, 6) and (0, −6), conjugate axis of length 14

6. vertices (0, 3) and (0, −3), conjugate axis of length 10

7. vertices (−2, 0) and (2, 0), conjugate axis of length 4

8. vertices (−3, 0) and (3, 0), foci (±5, 0)

9. vertices (0, 2) and (0, −2), foci (0, ±3)

10. vertices (0, −2) and (6, −2), foci $(3 \pm \sqrt{13}, -2)$

Graph each hyperbola. Identify the vertices, foci, and asymptotes.

11. $\frac{x^2}{9} - \frac{y^2}{36} = 1$

12. $\frac{y^2}{49} - \frac{x^2}{9} = 1$

13. $\frac{x^2}{16} - \frac{y^2}{1} = 1$

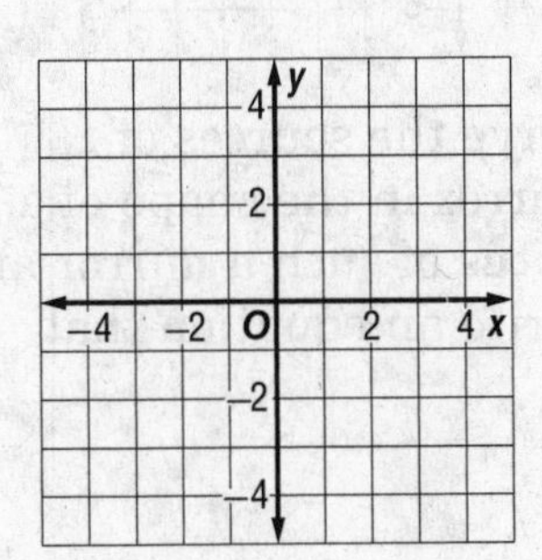

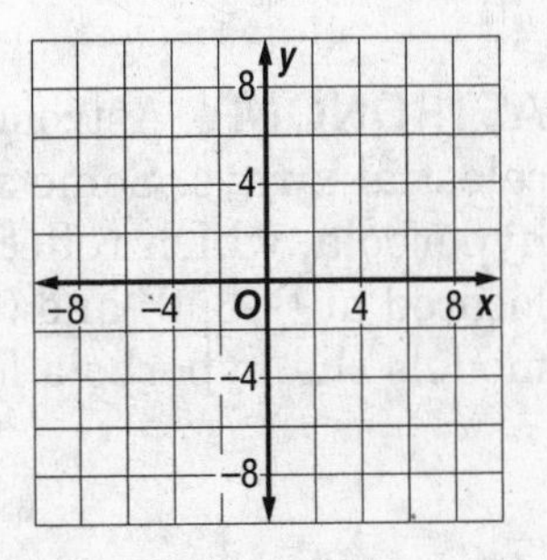

NAME ______________________ DATE __________ PERIOD __________

10-5 Practice

Hyperbolas

Write an equation for each hyperbola.

1. **2.** **3.**

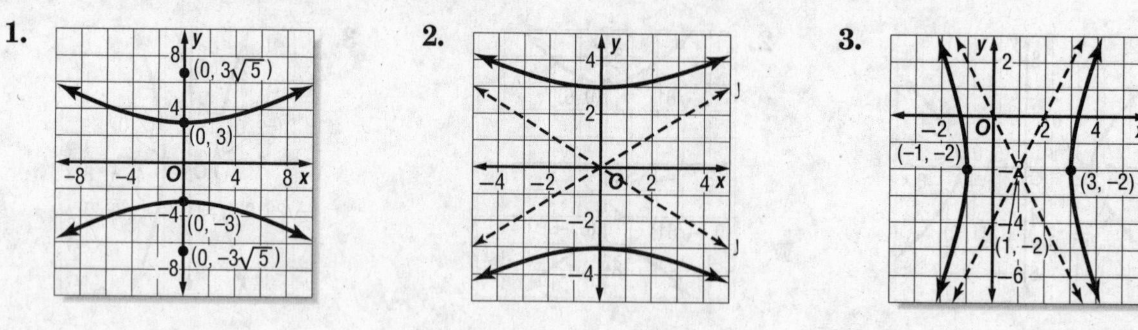

4. vertices (0, 7) and (0, −7), conjugate axis of length 18 units

5. vertices (0, −4) and (0, 4), conjugate axis of length 6 units

6. vertices (−5, 0) and (5, 0), foci $(\pm\sqrt{26}, 0)$

7. vertices (0, 2) and (0, −2), foci $(0, \pm\sqrt{5})$

Graph each hyperbola. Identify the vertices, foci, and asymptotes.

8. $\frac{y^2}{16} - \frac{x^2}{4} = 1$ **9.** $\frac{(y-2)^2}{1} - \frac{(x-1)^2}{4} = 1$ **10.** $\frac{(y+2)^2}{4} - \frac{(x-3)^2}{4} = 1$

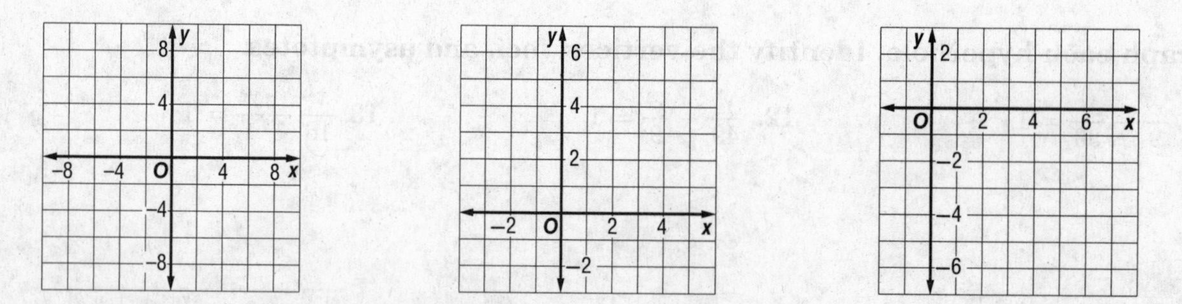

11. ASTRONOMY Astronomers use special x-ray telescopes to observe the sources of celestial x-rays. Some x-ray telescopes are fitted with a metal mirror in the shape of a hyperbola, which reflects the x-rays to a focus. Suppose the vertices of such a mirror are located at (−3, 0) and (3, 0), and one focus is located at (5, 0). Write an equation that models the hyperbola formed by the mirror.

NAME ______________________ DATE __________ PERIOD __________

10-6 Skills Practice

Identifying Conic Sections

Write each equation in standard form. State whether the graph of the equation is a *parabola, circle, ellipse,* or *hyperbola.* Then graph the equation.

1. $x^2 - 25y^2 = 25$

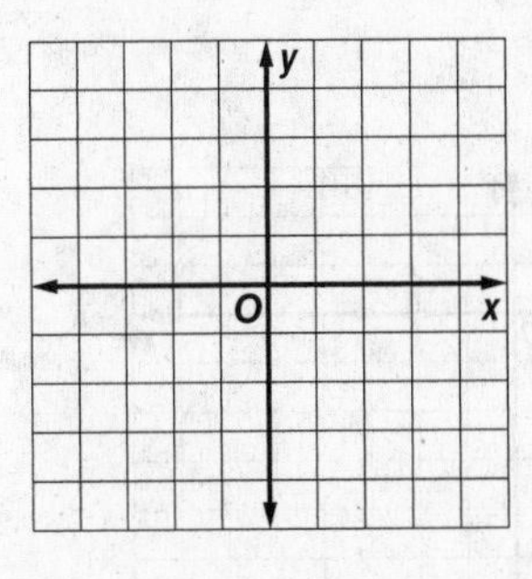

2. $9x^2 + 4y^2 = 36$

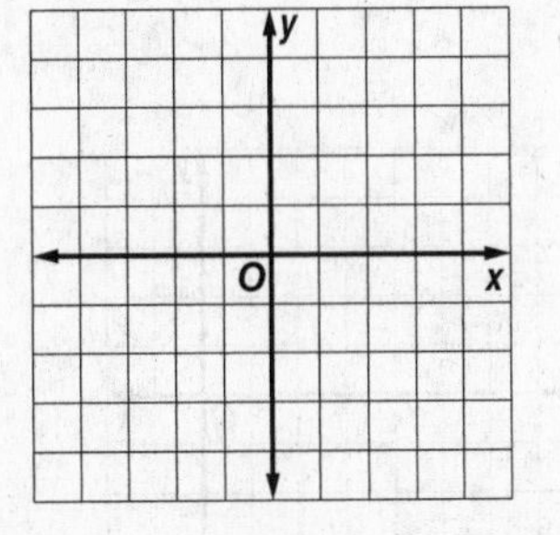

3. $x^2 + y^2 - 16 = 0$

4. $x^2 + 8x + y^2 = 9$

5. $x^2 + 2x - 15 = y$

6. $100x^2 + 25y^2 = 400$

Without writing the equation in standard form, state whether the graph of each equation is a *parabola, circle, ellipse,* or *hyperbola.*

7. $9x^2 + 4xy + 4y^2 = 36$

8. $x^2 + y^2 = 25$

9. $y = x^2 + 2x$

10. $8y^2 - 8xy = 2x^2 - 4x - 4$

11. $4y^2 - 25x^2 = 100$

12. $16x^2 + 5xy + y^2 = 16$

13. $16x^2 - 4y^2 = 64$

14. $5x^2 + 5y^2 = 25$

15. $25y^2 + 12xy + 9x^2 = 225$

16. $36y^2 - 4x^2 = 144$

17. $y = 4x^2 - 36x - 144$

18. $x^2 + y^2 - 144 = 0$

19. $(x + 3)^2 + (y - 1)^2 = 4$

20. $25y^2 - 50y + 4x^2 = 75$

21. $x^2 - 6y^2 + 9 = 0$

22. $x^2 - 2xy = y^2 + 5y - 6$

23. $(x + 5)^2 + y^2 = 10$

24. $25x^2 + 30xy + 10y^2 - 250 = 0$

NAME ______________________________ DATE ____________ PERIOD ____________

10-6 Practice

Identifying Conic Sections

Write each equation in standard form. State whether the graph of the equation is a *parabola, circle, ellipse,* or *hyperbola.* Then graph the equation.

1. $y^2 = -3x$

2. $x^2 + y^2 + 6x = 7$

3. $5x^2 - 6y^2 - 30x - 12y = -9$

4. $196y^2 = 1225 - 100x^2$

5. $3x^2 = 9 - 3y^2 - 6y$

6. $9x^2 + y^2 + 54x - 6y = -81$

Without writing the equation in standard form, state whether the graph of each equation is a *parabola, circle, ellipse,* or *hyperbola.*

7. $6x^2 + 6y^2 = 36$

8. $4x^2 + 6xy - y^2 = 16$

9. $9x^2 + 16y^2 - 10xy - 64y - 80 = 0$

10. $5x^2 + 5y^2 - 45 = 0$

11. $x^2 + 2x = y$

12. $4y^2 - 12xy - 36x^2 + 4x - 144 = 0$

13. ASTRONOMY A space probe flies by a planet in an hyperbolic orbit. It reaches the vertex of its orbit at (5, 0) and then travels along a path that gets closer and closer to the line $y = \frac{2}{5}x$. Write an equation that describes the path of the space probe if the center of its hyperbolic orbit is at (0, 0).

10-7 Skills Practice

Solving Linear-Nonlinear Systems

Solve each system of equations.

1. $y = x - 2$
$y = x^2 - 2$

2. $y = x + 3$
$y = 2x^2$

3. $y = 3x$
$x = y^2$

4. $y = x$
$x^2 + y^2 = 4$

5. $x = -5$
$x^2 + y^2 = 25$

6. $y = 7$
$x^2 + y^2 = 9$

7. $y = -2x + 2$
$y^2 = 2x$

8. $x - y + 1 = 0$
$y^2 = 4x$

9. $y = 2 - x$
$y = x^2 - 4x + 2$

10. $y = x - 1$
$y = x^2$

11. $y = 3x^2$
$y = -3x^2$

12. $y = x^2 + 1$
$y = -x^2 + 3$

13. $y = 4x$
$4x^2 + y^2 = 20$

14. $y = -1$
$4x^2 + y^2 = 1$

15. $4x^2 + 9y^2 = 36$
$x^2 - 9y^2 = 9$

16. $3(y + 2)^2 - 4(x - 3)^2 = 12$
$y = -2x + 2$

17. $x^2 - 4y^2 = 4$
$x^2 + y^2 = 4$

18. $y^2 - 4x^2 = 4$
$y = 2x$

Solve each system of inequalities by graphing.

19. $y \le 3x - 2$
$x^2 + y^2 < 16$

20. $y \le x$
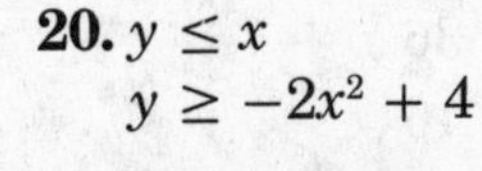
$y \ge -2x^2 + 4$

21. $4y^2 + 9x^2 < 144$
$x^2 + 8y^2 < 16$

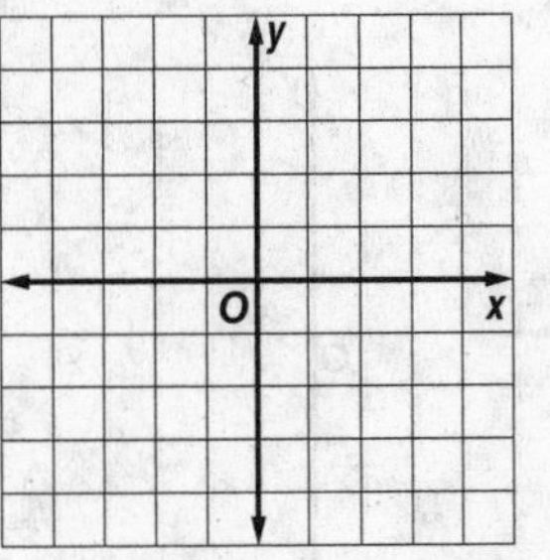

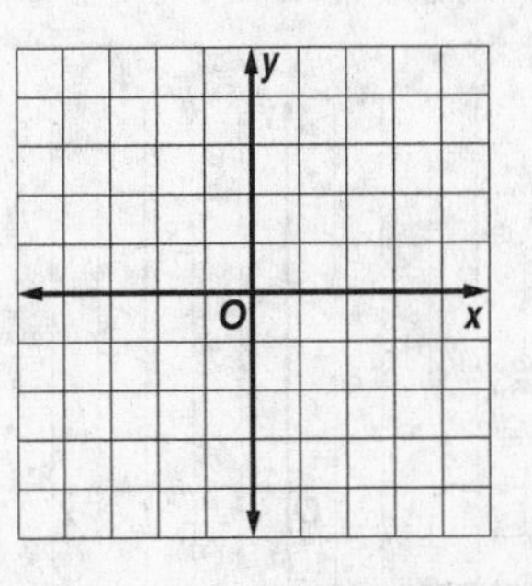

22. GARDENING An elliptical garden bed has a path from point A to point B. If the bed can be modeled by the equation $x^2 + 3y^2 = 12$ and the path can be modeled by the line $y = -\frac{1}{3}x$, what are the coordinates of points A and B?

NAME ______________________________ DATE ____________ PERIOD ____________

10-7 Practice

Solving Linear-Nonlinear Systems

Solve each system of equations.

1. $(x - 2)^2 + y^2 = 5$
$x - y = 1$

2. $x = 2(y + 1)^2 - 6$
$x + y = 3$

3. $y^2 - 3x^2 = 6$
$y = 2x - 1$

4. $x^2 + 2y^2 = 1$
$y = -x + 1$

5. $4y^2 - 9x^2 = 36$
$4x^2 - 9y^2 = 36$

6. $y = x^2 - 3$
$x^2 + y^2 = 9$

7. $x^2 + y^2 = 25$
$4y = 3x$

8. $y^2 = 10 - 6x^2$
$4y^2 = 40 - 2x^2$

9. $x^2 + y^2 = 25$
$x = 3y - 5$

10. $4x^2 + 9y^2 = 36$
$2x^2 - 9y^2 = 18$

11. $x = -(y - 3)^2 + 2$
$x = (y - 3)^2 + 3$

12. $\frac{x^2}{9} - \frac{y^2}{16} = 1$
$x^2 + y^2 = 9$

13. $25x^2 + 4y^2 = 100$
$x = -\frac{5}{2}$

14. $x^2 + y^2 = 4$
$\frac{x^2}{4} + \frac{y^2}{8} = 1$

15. $x^2 - y^2 = 3$
$y^2 - x^2 = 3$

16. $\frac{x^2}{7} + \frac{y^2}{7} = 1$
$3x^2 - y^2 = 9$

17. $x + 2y = 3$
$x^2 + y^2 = 9$

18. $x^2 + y^2 = 64$
$x^2 - y^2 = 8$

Solve each system of inequalities by graphing.

19. $y \geq x^2$
$y > -x + 2$

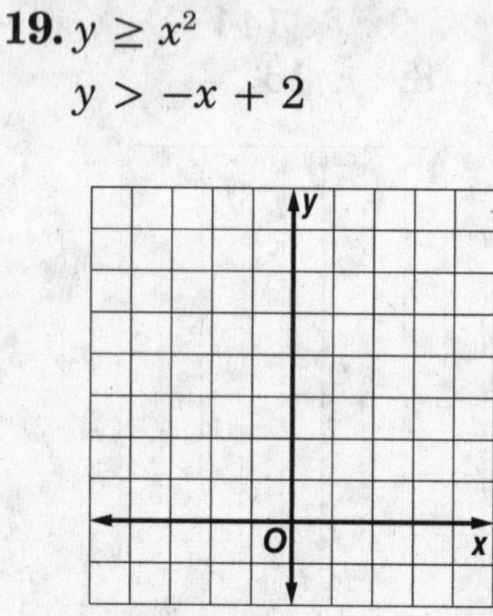

20. $x^2 + y^2 < 36$
$x^2 + y^2 \geq 16$

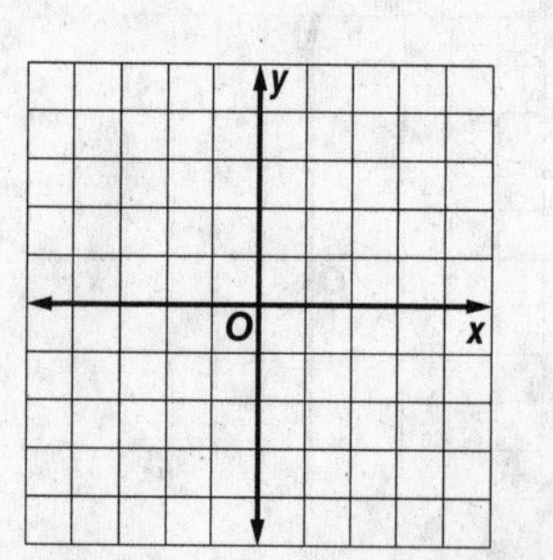

21. $\frac{(y - 3)^2}{16} + \frac{(x + 2)^2}{4} \leq 1$
$(x + 1)^2 + (y - 2)^2 \leq 4$

22. GEOMETRY The top of an iron gate is shaped like half an ellipse with two congruent segments from the center of the ellipse to the ellipse as shown. Assume that the center of the ellipse is at (0, 0). If the ellipse can be modeled by the equation $x^2 + 4y^2 = 4$ for $y \geq 0$ and the two congruent segments can be modeled by $y = \frac{\sqrt{3}}{2}x$ and $y = -\frac{\sqrt{3}}{2}x$, what are the coordinates of points A and B?

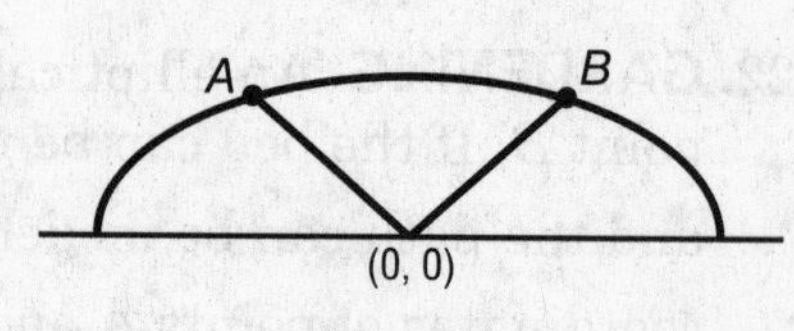

NAME ______________________ DATE ____________ PERIOD ____________

11-1 Skills Practice

Sequences as Functions

Find the next four terms of each arithmetic sequence. Then graph the sequence.

1. 7, 11, 15, …

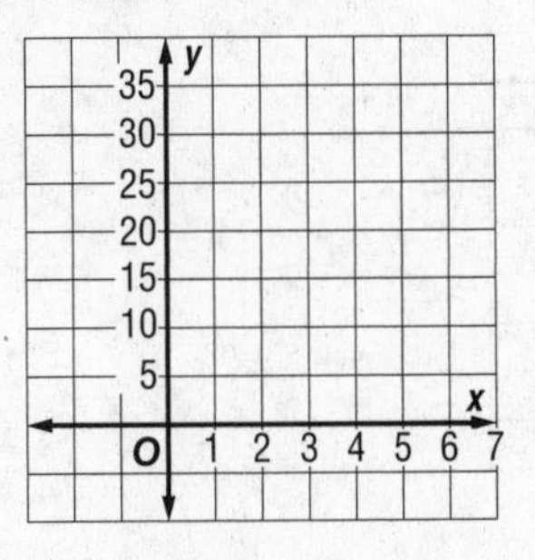

2. −10, −5, 0, …

3. 101, 202, 303, …

4. 15, 7, −1, …

Find the next three terms for each geometric sequence. Then graph the sequence.

5. $\frac{1}{2}$, 2, 8, …

6. $\frac{2}{5}$, 2, 10, …

7. $6\frac{1}{3}$, 19, 57, …

8. 13, 26, 52, …

11-1 Practice

Sequences as Functions

Find the next four terms of each arithmetic sequence. Then graph the sequence.

1. 5, 8, 11, ...

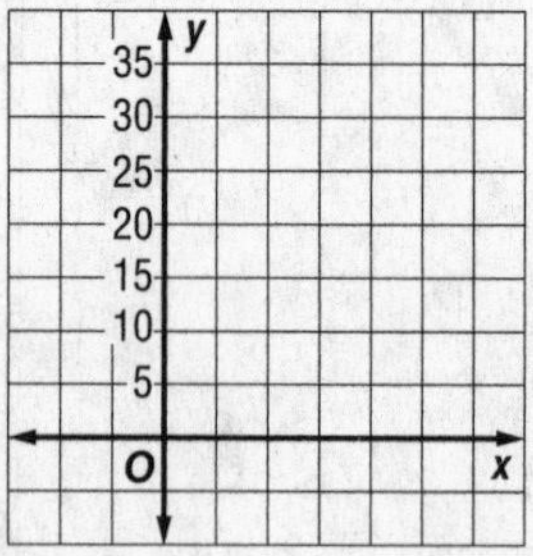

2. –4, –6, –8, ...

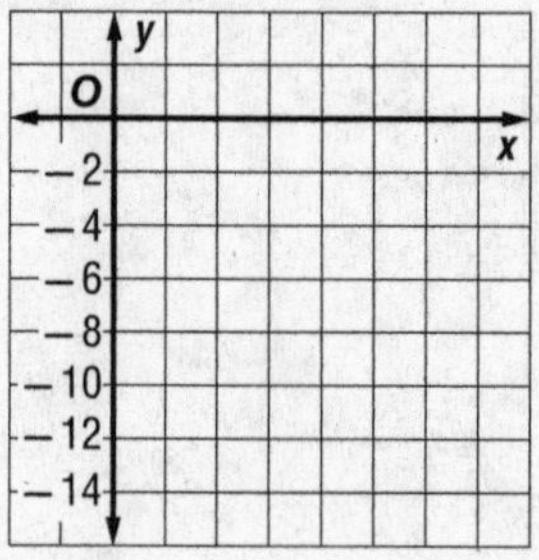

Find the next three terms for each geometric sequence. Then graph the sequence.

3. $\frac{1}{10}, \frac{1}{2}, 2\frac{1}{2}, \ldots$

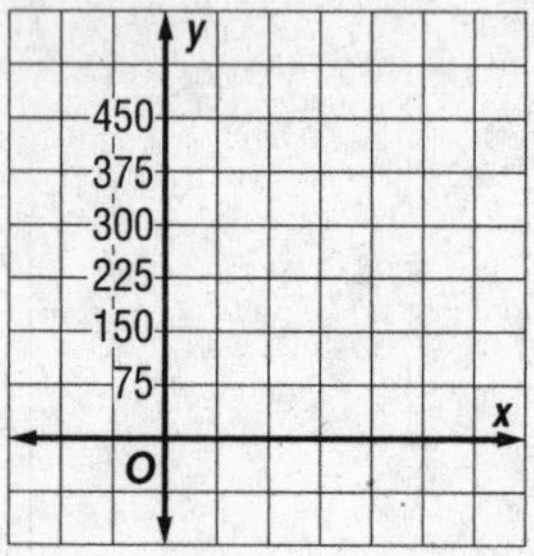

4. 81, 27, 9, ...

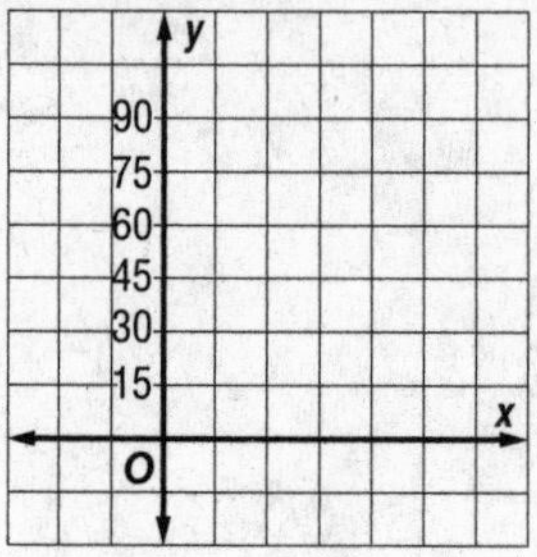

Determine whether each sequence is *arithmetic*, *geometric*, or *neither*. Explain your reasoning.

5. 57, 456, 3648, 29,184 , ...

6. –49, –37, –25, –13, ...

7. 4, 9, 16, 25, 36, ...

8. 824, 412, 206, 103, ...

9. EDUCATION Trevor Koba has opened an English Language School in Isehara, Japan. He began with 26 students. If he enrolls 3 new students each week, in how many weeks will he have 101 students?

10. SALARIES Yolanda interviewed for a job that promised her a starting salary of $32,000 with a $1250 raise at the end of each year. What will her salary be during her sixth year if she accepts the job?

NAME ______________________ DATE ____________ PERIOD ________

11-2 Skills Practice

Arithmetic Sequences and Series

Find the indicated term of each arithmetic sequence.

1. $a_1 = 56, d = 13, n = 73$

2. a_{19} for 16, 32, 48, …

Write an equation for the *n*th term of each arithmetic sequence.

3. 64, 78, 92, 106, …

4. −416, −323, −230, −137, …

Find the arithmetic means in each sequence.

5. 17, $\underline{\ ?\ }$, $\underline{\ ?\ }$, $\underline{\ ?\ }$, 41

6. 235, $\underline{\ ?\ }$, $\underline{\ ?\ }$, $\underline{\ ?\ }$, $\underline{\ ?\ }$, $\underline{\ ?\ }$, $\underline{\ ?\ }$, 32

Find the sum of each arithmetic series.

7. $1 + 4 + 7 + 10 + \ldots + 43$

8. $5 + 8 + 11 + 14 + \ldots + 32$

9. $3 + 5 + 7 + 9 + \ldots + 19$

10. $-2 + (-5) + (-8) + \ldots + (-20)$

11. $\sum_{n=1}^{5}(2n - 3)$

12. $\sum_{n=1}^{18}(10 + 3n)$

13. $\sum_{n=2}^{10}(4n + 1)$

14. $\sum_{n=5}^{12}(4 - 3n)$

Find the first three terms of each arithmetic series.

15. $a_1 = 4, a_n = 31, S_n = 175$

16. $a_1 = -3, a_n = 41, S_n = 228$

17. $n = 10, a_n = 41, S_n = 230$

18. $n = 19, a_n = 85, S_n = 760$

11-2 Practice

Arithmetic Sequences and Series

Find the indicated term of each arithmetic sequence.

1. Find the sixtieth term of the arithmetic sequence if $a_1 = 418$ and $d = 12$.

2. Find a_{23} in the sequence, $-18, -34, -50, -66, \ldots$.

Write an equation for the *n*th term of each arithmetic sequence.

3. 45, 30, 15, 0, ...

4. −87, −73, −59, −45, ...

Find the sum of each arithmetic series.

5. $5 + 7 + 9 + 11 + \ldots + 27$

6. $-4 + 1 + 6 + 11 + \ldots + 91$

7. $13 + 20 + 27 + \ldots + 272$

8. $89 + 86 + 83 + 80 + \ldots + 20$

9. $\sum_{n=1}^{4}(1 - 2n)$

10. $\sum_{j=1}^{6}(5 + 3n)$

11. $\sum_{n=1}^{5}(9 - 4n)$

12. $\sum_{k=4}^{10}(2k + 1)$

13. $\sum_{n=3}^{8}(5n - 10)$

14. $\sum_{n=1}^{101}(4 - 4n)$

Find the first three terms of each arithmetic series described.

15. $a_1 = 14$, $a_n = -85$, $S_n = -1207$

16. $a_1 = 1$, $a_n = 19$, $S_n = 100$

17. $n = 16$, $a_n = 15$, $S_n = -120$

18. $n = 15$, $a_n = 5\frac{4}{5}$, $S_n = 45$

19. STACKING A health club rolls its towels and stacks them in layers on a shelf. Each layer of towels has one less towel than the layer below it. If there are 20 towels on the bottom layer and one towel on the top layer, how many towels are stacked on the shelf?

20. BUSINESS A merchant places $1 in a jackpot on August 1, then draws the name of a regular customer. If the customer is present, he or she wins the $1 in the jackpot. If the customer is not present, the merchant adds $2 to the jackpot on August 2 and draws another name. Each day the merchant adds an amount equal to the day of the month. If the first person to win the jackpot wins $496, on what day of the month was her or his name drawn?

NAME ______________________ DATE __________ PERIOD __________

11-3 Skills Practice

Geometric Sequences and Series

Find a_n for each geometric sequence.

1. $a_1 = 5, r = 2, n = 6$

2. $a_1 = 18, r = 3, n = 6$

3. $a_1 = -3, r = -2, n = 5$

4. $a_1 = -20, r = -2, n = 9$

5. $a_1 = 65{,}536, r = \frac{1}{4}, n = 6$

6. $a_1 = -78{,}125, r = \frac{1}{5}, n = 9$

Write an equation for the *n*th term of each geometric sequence.

7. 3, 9, 27, ...

8. −1, −3, −9, ...

9. 2, −6, 18, ...

10. 5, 10, 20, ...

11. 12, 36, 108, 324, ...

12. 32,768; 4096; 512; 64; ...

13. 25, 175, 1225, 8575, ...

14. −16,384; −8192; −4096; −2048; ...

Find the geometric means of each sequence.

15. 4, ?, ?, ?, 64

16. 1, ?, ?, ?, 81

17. 38; 228; ?; 8208; 49,248; ...

18. 51; ?; 4131; ?; 334,611; ...

19. −15, ?, ?, ?, −240, ...

20. 531,441; ?; ?; ?; ?; 9; ...

Find a_1 for each geometric series described.

21. $S_n = 1295, r = 6, n = 4$

22. $S_n = 1640, r = 3, n = 8$

23. $S_n = 218\frac{2}{5}, a_n = 1\frac{2}{5}, r = \frac{1}{5}$

24. $S_n = -342, a_n = -512, r = -2$

NAME ______________________ DATE __________ PERIOD __________

11-3 Practice

Geometric Sequences and Series

Find a_n for each geometric sequence.

1. $a_1 = 5, r = 3, n = 6$
2. $a_1 = 20, r = -3, n = 6$
3. $a_1 = -4, r = -2, n = 10$
4. a_8 for $-\frac{1}{250}, -\frac{1}{50}, -\frac{1}{10}, \ldots$
5. a_{12} for 96, 48, 24, …
6. $a_1 = 8, r = \frac{1}{2}, n = 9$
7. $a_1 = -3125, r = -\frac{1}{5}, n = 9$
8. $a_1 = 3, r = \frac{1}{10}, n = 8$

Write an equation for the nth term of each geometric sequence.

9. 1, 4, 16, …
10. −1, −5, −25, …
11. $1, \frac{1}{2}, \frac{1}{4}, \ldots$
12. −3, −6, −12, …

Find the sum of each geometric series.

13. $\sum_{k=3}^{10}(-4)(-2)^{k-1}$
14. $\sum_{k=1}^{8}(-3)(3)^{k-1}$
15. $\sum_{k=2}^{32}9(-1)^{k-1}$

Find a_1 for each geometric series described.

16. $S_n = 1550, n = 3, r = 5$
17. $S_n = 1512, n = 6, r = 2$
18. $S_n = 3478.2, r = 2, a_n = 3481.6$
19. $S_n = 4860, r = 3, a_n = 3280.5$

20. **BIOLOGY** A culture initially contains 200 bacteria. If the number of bacteria doubles every 2 hours, how many bacteria will be in the culture at the end of 12 hours?

21. **LIGHT** If each foot of water in a lake screens out 60% of the light above, what percent of the light passes through 5 feet of water?

22. **INVESTING** Raul invests $1000 in a savings account that earns 5% interest compounded annually. How much money will he have in the account at the end of 5 years?

NAME ______________________ DATE __________ PERIOD ________

11-4 Skills Practice

Infinite Geometric Series

Find the sum of each infinite series, if it exists.

1. $a_1 = 1, r = \frac{1}{2}$

2. $a_1 = 5, r = -\frac{2}{5}$

3. $a_1 = 8, r = 2$

4. $a_1 = 6, r = \frac{1}{2}$

5. $4 + 2 + 1 + \frac{1}{2} + \ldots$

6. $540 - 180 + 60 - 20 + \ldots$

7. $5 + 10 + 20 + \ldots$

8. $-336 + 84 - 21 + \ldots$

9. $125 + 25 + 5 + \ldots$

10. $9 - 1 + \frac{1}{9} - \ldots$

11. $\frac{3}{4} + \frac{9}{4} + \frac{27}{4} + \ldots$

12. $\frac{1}{3} + \frac{1}{9} + \frac{1}{27} + \ldots$

13. $5 + 2 + 0.8 + \ldots$

14. $9 + 6 + 4 + \ldots$

15. $\sum_{n=1}^{\infty} 10\left(\frac{1}{2}\right)^{n-1}$

16. $\sum_{n=1}^{\infty} 6\left(-\frac{1}{3}\right)^{n-1}$

17. $\sum_{n=1}^{\infty} 15\left(\frac{2}{5}\right)^{n-1}$

18. $\sum_{n=1}^{\infty} \left(-\frac{4}{3}\right)\left(\frac{1}{3}\right)^{n-1}$

Write each repeating decimal as a fraction.

19. $0.\overline{4}$

20. $0.\overline{8}$

21. $0.\overline{27}$

22. $0.\overline{67}$

23. $0.\overline{54}$

24. $0.\overline{375}$

25. $0.\overline{641}$

26. $0.\overline{171}$

NAME ______________________________ DATE ____________ PERIOD ____________

11-4 Practice

Infinite Geometric Series

Find the sum of each infinite series, if it exists.

1. $a_1 = 35, r = \frac{2}{7}$

2. $a_1 = 26, r = \frac{1}{2}$

3. $a_1 = 98, r = -\frac{3}{4}$

4. $a_1 = 42, r = \frac{6}{5}$

5. $a_1 = 112, r = -\frac{3}{5}$

6. $a_1 = 500, r = \frac{1}{5}$

7. $a_1 = 135, r = -\frac{1}{2}$

8. $18 - 6 + 2 - \ldots$

9. $2 + 6 + 18 + \ldots$

10. $6 + 4 + \frac{8}{3} + \ldots$

11. $\frac{4}{25} + \frac{2}{5} + 1 + \ldots$

12. $10 + 1 + 0.1 + \ldots$

13. $100 + 20 + 4 + \ldots$

14. $-270 + 135 - 67.5 + \ldots$

15. $0.5 + 0.25 + 0.125 + \ldots$

16. $\frac{7}{10} + \frac{7}{100} + \frac{7}{1000} + \ldots$

17. $0.8 + 0.08 + 0.008 + \ldots$

18. $\frac{1}{12} - \frac{1}{6} + \frac{1}{3} - \ldots$

19. $3 + \frac{9}{7} + \frac{27}{49} + \ldots$

20. $0.3 - 0.003 + 0.00003 - \ldots$

21. $0.06 + 0.006 + 0.0006 + \ldots$

22. $\frac{2}{3} - 2 + 6 - \ldots$

23. $\sum_{n=1}^{\infty} 3\left(\frac{1}{4}\right)^{n-1}$

24. $\sum_{n=1}^{\infty} \frac{2}{3}\left(-\frac{3}{4}\right)^{n-1}$

25. $\sum_{n=1}^{\infty} 18\left(\frac{2}{3}\right)^{n-1}$

26. $\sum_{n=1}^{\infty} 5(-0.1)^{n-1}$

Write each repeating decimal as a fraction.

27. $0.\overline{6}$

28. $0.\overline{09}$

29. $0.\overline{43}$

30. $0.\overline{27}$

31. $0.\overline{243}$

32. $0.8\overline{4}$

33. $0.\overline{990}$

34. $0.\overline{150}$

35. PENDULUMS On its first swing, a pendulum travels 8 feet. On each successive swing, the pendulum travels $\frac{4}{5}$ the distance of its previous swing. What is the total distance traveled by the pendulum when it stops swinging?

36. ELASTICITY A ball dropped from a height of 10 feet bounces back $\frac{9}{10}$ of that distance. With each successive bounce, the ball continues to reach $\frac{9}{10}$ of its previous height. What is the total vertical distance (both up and down) traveled by the ball when it stops bouncing? (*Hint*: Add the total distance the ball falls to the total distance it rises.)

11-5 Skills Practice

Recursion and Iteration

Find the first five terms of each sequence described.

1. $a_1 = 4, a_{n+1} = a_n + 7$

2. $a_1 = -2, a_{n+1} = a_n + 3$

3. $a_1 = 5, a_{n+1} = 2a_n$

4. $a_1 = -4, a_{n+1} = 6 - a_n$

5. $a_1 = 0, a_2 = 1, a_{n+1} = a_n + a_{n-1}$

6. $a_1 = -1, a_2 = -1, a_{n+1} = a_n - a_{n-1}$

7. $a_1 = 3, a_2 = -5, a_{n+1} = -4a_n + a_{n-1}$

8. $a_1 = -3, a_2 = 2, a_{n+1} = a_{n-1} - a_n$

Write a recursive formula for each sequence.

9. 2, 7, 12, 17, 22, ...

10. 4; 16; 256; 65,536, ...

11. 1, 1, 2, 3, 5, 8, ...

12. 1; 3; 11; 123; 15,131, ...

13. 2, 3, 6, 18, 108, 1944, ...

14. –2, –8, –512, –134, 217, 728, ...

Find the first three iterates of each function for the given initial value.

15. $f(x) = 2x - 1, x_0 = 3$

16. $f(x) = 5x - 3, x_0 = 2$

17. $f(x) = 3x + 4, x_0 = -1$

18. $f(x) = 4x + 7, x_0 = -5$

19. $f(x) = -x - 3, x_0 = 10$

20. $f(x) = -3x + 6, x_0 = 6$

21. $f(x) = -3x + 4, x_0 = 2$

22. $f(x) = 6x - 5, x_0 = 1$

23. $f(x) = 7x + 1, x_0 = -4$

24. $f(x) = x^2 - 3x, x_0 = 5$

NAME ______________________ DATE ____________ PERIOD ________

11-5 Practice

Recursion and Iteration

Find the first five terms of each sequence described.

1. $a_1 = 3, a_{n+1} = a_n + 5$

2. $a_1 = -7, a_{n+1} = a_n + 8$

3. $a_1 = -3, a_{n+1} = 3a_n + 2$

4. $a_1 = -8, a_{n+1} = 10 - a_n$

5. $a_1 = 2, a_2 = -3, a_{n+1} = 5a_n - 8a_{n-1}$

6. $a_1 = -2, a_2 = 1, a_{n+1} = -2a_n + 6a_{n-1}$

Write a recursive formula for each sequence.

7. 2, 5, 7, 12, 19, ...

8. 1, –2, –2, 4, –8, –32, 256, ...

9. –3, 9, 81, 6561, ...

10. 3, 7, 4, –3, –7, –4, 3, 7, ...

11. –1, 2, 1, 3, 4, 7, 11, ...

12. 3, 1, $\frac{1}{3}$, $\frac{1}{3}$, 1, 3, 3, 1, ...

Find the first three iterates of each function for the given initial value.

13. $f(x) = 3x + 4, x_0 = -1$

14. $f(x) = 10x + 2, x_0 = -1$

15. $f(x) = 8 + 3x, x_0 = 1$

16. $f(x) = 8 - x, x_0 = -3$

17. $f(x) = 4x + 5, x_0 = -1$

18. $f(x) = 5(x + 3), x_0 = -2$

19. $f(x) = -8x + 9, x_0 = 1$

20. $f(x) = -4x^2, x_0 = -1$

21. $f(x) = x^2 - 1, x_0 = 3$

22. $f(x) = 2x^2; x_0 = 5$

23. INFLATION Iterating the function $c(x) = 1.05x$ gives the future cost of an item at a constant 5% inflation rate. Find the cost of a $2000 ring in five years at 5% inflation.

24. FRACTALS Replacing each side of the square shown with the combination of segments below it gives the figure to its right.

3 in.

1 in.
1 in. 1 in.
1 in. 1 in.

a. What is the perimeter of the original square?

b. What is the perimeter of the new shape?

c. If you repeat the process by replacing each side of the new shape by a proportional combination of 5 segments, what will the perimeter of the third shape be?

d. What function $f(x)$ can you iterate to find the perimeter of each successive shape if you continue this process?

NAME ______________________ DATE ____________ PERIOD ________

11-6 Skills Practice

The Binomial Theorem

Expand each binomial.

1. $(x - y)^3$

2. $(a + b)^5$

3. $(g - h)^4$

4. $(m + 1)^4$

5. $(r + 4)^3$

6. $(a - 5)^4$

7. $(y - 7)^3$

8. $(d + 2)^5$

9. $(x - 1)^4$

10. $(2a + b)^4$

11. $(c - 4d)^3$

12. $(2a + 3)^3$

Find the indicated term of each expression.

13. fourth term of $(6x + 5)^5$

14. fifth term of $(x - 3y)^6$

15. third term of $(11x + 3y)^6$

16. twelfth term of $(13x - 4y)^{11}$

17. fourth term of $(m + n)^{10}$

18. seventh term of $(x - y)^8$

19. third term of $(b + 6)^5$

20. sixth term of $(r - 2)^9$

21. fifth term of $(2a + 3)^6$

22. second term of $(3x - y)^7$

11-6 Practice

The Binomial Theorem

Expand each binomial.

1. $(n + v)^5$

2. $(x - y)^4$

3. $(x + y)^6$

4. $(r + 3)^5$

5. $(m - 5)^5$

6. $(x + 4)^4$

7. $(3x + y)^4$

8. $(2m - y)^4$

9. $(w - 3z)^3$

10. $(2d + 3)^6$

11. $(x + 2y)^5$

12. $(2x - y)^5$

13. $(a - 3b)^4$

14. $(3 - 2z)^4$

15. $(3m - 4p)^3$

16. $(5x - 2y)^4$

Find the indicated term of each expansion.

17. sixth term of $(x + 4y)^6$

18. fourth term of $(5x + 2y)^5$

19. eighth term of $(x - y)^{11}$

20. third term of $(x - 2)^8$

21. seventh term of $(a + b)^{10}$

22. sixth term of $(m - p)^{10}$

23. ninth term of $(r - t)^{14}$

24. tenth term of $(2x + y)^{12}$

25. fourth term of $(x - 3y)^6$

26. fifth term of $(2x - 1)^9$

27. GEOMETRY How many line segments can be drawn between ten points, no three of which are collinear, if you use exactly two of the ten points to draw each segment?

28. PROBABILITY If you toss a coin 4 times, how many different sequences of tosses will give exactly 3 heads and 1 tail or exactly 1 head and 3 tails?

NAME ______________________ DATE ____________ PERIOD ________

11-7 Skills Practice

Proof by Mathematical Induction

Prove that each statement is true for all natural numbers.

1. $1 + 3 + 5 + \ldots + (2n - 1) = n^2$

2. $2 + 4 + 6 + \ldots + 2n = n^2 + n$

3. $6^n - 1$ is divisible by 5.

Find a counterexample to disprove each statement.

4. $n^2 - n + 11$ is prime.

5. $1 + 4 + 8 + \ldots + 2n = \dfrac{n(n + 1)(2n + 1)}{6}$

11-7 Practice

Proof by Mathematical Induction

Prove that each statement is true for all natural numbers.

1. $1 + 2 + 4 + 8 + \ldots + 2^{n-1} = 2^n - 1$

2. $1 + 4 + 9 + \ldots + n^2 = \frac{n(n + 1)(2n + 1)}{6}$

3. $18^n - 1$ is a multiple of 17.

Find a counterexample to disprove each statement.

4. $1 + 4 + 7 + \ldots + (3n - 2) = n^3 - n^2 + 1$

5. $5^n - 2n - 3$ is divisible by 3.

6. $1 + 3 + 5 + \ldots + (2n - 1) = \frac{n^2 + 3n - 2}{2}$

7. $1^3 + 2^3 + 3^3 + \ldots + n^3 = n^4 - n^3 + 1$

12-1 Skills Practice

Experiments, Surveys, and Observational Studies

Exercises

State whether each situation represents an *experiment* or an *observational study.* If it is an experiment, then identify the control group and treatment group. Then determine whether there is bias.

1. Find 200 people at a mall and randomly split them into two groups. One group tries a new pain reliever medicine, and the other group tries a placebo.

2. Find 100 students, half of whom play sports, and compare their SAT scores.

Determine whether the following situation calls for a *survey,* an *observational study,* or an *experiment.* Explain the process.

3. You want to find opinions on the best computer game to buy.

4. You want to see if students who have a 4.0 grade point average study more than those who do not.

Determine whether the following statements show *correlation* or *causation.* Explain.

5. When a traffic light is red, a driver brings her car to a stop.

6. Studies have shown that students who are confident before a test raise their test scores.

7. If I practice the saxophone every day, I will make the school jazz band.

8. If water is heated to 100° Celsius, it will boil.

12-1 Practice

Experiments, Surveys, and Observational Studies

State whether each situation represents an *experiment* or an *observational study*. If it is an experiment, then identify the control group and treatment group. Then determine whether there is bias.

1. Find 300 students, half of whom are on the chess team, and compare their grade point averages.

2. Find 1000 people and randomly split them into two groups. Give a new vitamin to one group and a placebo to the other group.

Determine whether each situation call for a *survey*, an *observational study*, or an *experiment*. Explain the process.

3. You want to compare the health of students who walk to school to the health of students who ride the bus.

4. You want to find out if people who eat a candy bar immediately before a math test get higher scores than people who do not.

Determine whether the following statements show *correlation* or *causation*. Explain.

5. If I jog every day, I can complete a marathon in three hours.

6. When there are no clouds in the sky, it does not rain.

7. Studies show that taking a multivitamin leads to a longer life.

8. If I study for three hours, I will earn a grade of 100% on my history test.

NAME ______________________ DATE ____________ PERIOD ________

12-2 Skills Practice

Statistical Analysis

Which measure of central tendency best represents the data, and why?

1. {10.2, 11.5, 299.7, 15.5, 20, 8.4, 2.4, 12.7}

2. {75, 60, 60, 71, 74.5, 60, 67, 72.5}

3. {200, 250, 225, 25, 268, 250, 275, 238, 259}

4. {410, 405, 397, 450, 376, 422, 401}

Determine whether the following represents a *population* or a *sample*.

5. a school lunch survey that asks every fifth student that enters the lunch room

6. Tenth graders at a high school are surveyed about school athletics.

7. a list of the test scores of all the students in a class

8. a list of the scores of 1000 students on an SAT test

9. MOVIES A survey of 728 random people found that 72% prefer comedies over romantic movies. What is the margin of sampling error and the likely interval that contains the percent of the population?

10. SPORTS A survey of 3441 random people in one U.S. state found that 80% watched College football games every weekend in the Fall. What is the margin of sampling error and the likely interval that contains the percent of the population?

11. Determine whether each is a sample or a population. Then find the standard deviation of the data. Round to the nearest hundredth.

a.

The Shoe Sizes of 12 Students at a High School					
4	8	5	6	6	5
9	7	10	7	9	8

b.

The Number of Sit Ups Completed by All Students in a Gym Class					
50	28	41	61	54	28
47	33	45	50	50	61
23	41	31	38	42	

NAME ______________________________ DATE ____________ PERIOD ____________

12-2 Practice

Statistical Analysis

Which measure of central tendency best represents the data, and why?

1. {12.1, 14.9, 6.7, 10, 12.8, 14, 18}

2. {77.9, 101, 78.9, 105, 4.2, 110, 87.9}

3. {10, 14.7, 14.7, 21, 7.4, 14.7, 8, 14.7}

4. {29, 36, 14, 99, 16, 15, 12, 30}

Determine whether the following represents a *population* or a *sample*.

5. a list of the times every student in gym class took to run a mile

6. the test scores of seven students in a chemistry class are compared

7. friends compare the batting averages of players who are listed in their collections of baseball cards

8. every student in a high school votes in a class president election

9. CARS A survey of 56 random people in a small town found that 14% drive convertibles year-round. What is the margin of sampling error? What is the likely interval that contains the percentage of the population that drives convertibles year-round?

10. BEACHES A survey of 812 random people in Hawaii found that 57% went to the beach at least four times last July. What is the margin of sampling error? What is the likely interval that contains the percentage of the population that went to the beach at least four times last July?

11. Determine whether each is a sample or a population. Then find the standard deviation of the data. Round to the nearest hundredth.

a.

The Number of Wins for Each Player on a Tennis Team Last Season					
10	2	9	17	4	8
9	9	10	15	19	5

b.

The Number of Medals Earned by 18 High School Debaters					
7	10	4	5	10	9
11	5	6	4	4	3
12	7	8	5	3	9

NAME ______________________ DATE ____________ PERIOD ________

12-3 Skills Practice

Conditional Probability

The spinner is numbered from one to eight. Find each probability.

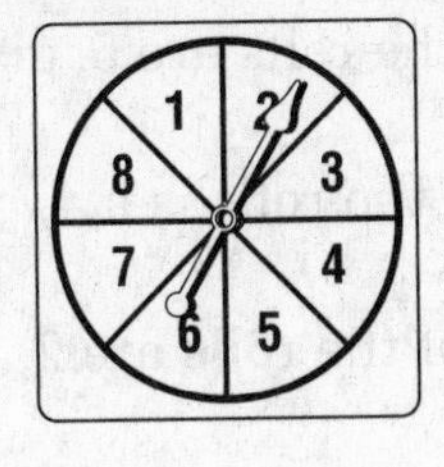

1. The spinner lands on 3, given that the spinner lands on an odd number.

2. The spinner lands on a number greater than 5, given that the spinner lands on an even number.

3. The spinner lands on a number less than 6, given that the spinner does not land on 1 or 2.

4. The spinner lands on 7, given that the spinner lands on a number greater than 4.

5. The spinner does not land on 3, given that the spinner lands on an odd number.

6. The spinner lands on a number less than 6, given that the spinner lands on an odd number.

7. CONCERTS The Panic Squadron is playing a concert. Jennifer surveyed her classmates to see if they were in the Panic Squadron Fan Club and if they were going to a concert. Find the probability that a person surveyed went to the concert, given that they are a fan club member.

	Fan Club Member	Not in Fan Club
Going to Concert	12	4
Not Going to Concert	2	18

8. SPECIAL ELECTIONS When a U.S. congressman vacates their office in the middle of their two-year term, a special election for the remainder of the term is often held to fill the vacancy. The table shows the number of special elections won by each party between 2004 and 2007. Find each probability.

	2004	2005	2006	2007
Republican Wins	0	1	2	3
Democrat Wins	3	1	1	2

Source: Clerk of the House of Representatives

a. A Republican wins, given the special election was held in 2007.

b. The election was held in 2005, given the winner was a Democrat.

c. A Democrat wins, given the election was held in 2006.

NAME ______________________ DATE __________ PERIOD __________

12-3 Practice

Conditional Probability

Four dice are thrown. Find each probability.

1. All of the rolls are 5, given that all of the rolls show the same number.

2. One of the rolls is a 4, given that all of the rolls are greater than 3.

3. None of the rolls are 2, given that all of the rolls are even.

4. One of the rolls is a 6, given that one of the rolls is a 2.

5. CHEMISTRY Cheryl and Jerome are testing the pH of 32 unknown substances as part of science class. Cheryl and Jerome split the work between them as shown in the table. Find each probability.

Results	Cheryl's Tests	Jerome's Tests
Acidic	12	8
Basic	9	3

a. The substance is acidic, given that Cheryl is testing it.

b. Jerome is testing the substance, given that it is basic.

6. ELECTIONS Rose Heck is running against Joe Coniglio in a district that includes the towns of Hasbrouck, Clinton, Eastwick, and Abletown. The table shows how many votes each candidate received in each town. Find each probability.

	Hasbrouck	Clinton	Eastwick	Abletown
Joe Coniglio	1743	1782	886	7790
Rose Heck	2616	2178	1329	5876

a. A voter cast a ballot for Joe Coniglio, given that the voter cast a ballot in Clinton.

b. A voter cast a ballot for Rose Heck, given that the voter cast a ballot in Eastwick.

c. A voter cast a ballot in Hasbrouck, given that the voter cast a ballot for Joe Coniglio.

7. BASEBALL Derek Jeter, a player for the New York Yankees, had 206 hits in the 2007 Major League Baseball season and has 2356 career hits. The table below shows the number of singles, doubles, triples, and home runs Derek Jeter had in the 2007 season and during his career. Find each probability.

	Singles	Doubles	Triples	Home Runs
2007 Season	151	39	4	12
Career	1721	386	54	195

a. A hit was a home run, given that the hit happened in the 2007 season.

b. A hit was a double, given that the hit happened during Jeter's career.

NAME ______________________ DATE ____________ PERIOD ____________

12-4 Skills Practice

Probability and Probability Distributions

1. **PHOTOGRAPHY** Ahmed is posting 2 photographs on his website. He has narrowed his choices to 4 landscape photographs and 3 portraits. If he chooses the two photographs at random, find the probability of each selection.

a. P(2 portrait) **b.** P(2 landscape) **c.** P(1 of each)

2. **VIDEOS** The Carubas have a collection of 28 movies, including 12 westerns and 16 science fiction. Elise selects 3 of the movies at random to bring to a sleep-over at her friend's house. Find the probability of each selection.

a. P(3 westerns) **b.** P(3 science fiction)

c. P(1 western and 2 science fiction) **d.** P(2 westerns and 1 science fiction)

e. P(3 comedy) **f.** P(2 science fiction and 2 westerns)

3. **CLASS** The chart to the right shows the class and gender statistics for the students taking an Algebra 1 or Algebra 2 class at La Mesa High School. If a student taking Algebra 1 or Algebra 2 is selected at random, find each probability. Express as decimals rounded to the nearest thousandth.

Class/Gender	Number
Freshman/Male	95
Freshman/Female	101
Sophomore/Male	154
Sophomore/Female	145
Junior/Male	100
Junior/Female	102

a. P(sophomore/female)

b. P(junior/male)

c. P(freshman/male)

d. P(freshman/female)

4. **FAMILY** Lisa has 10 cousins. Four of her cousins are older than her; six are younger. Seven of her cousins are boys, and 3 are girls. Find the probability of each when one cousin is chosen at random.

a. P(girl) **b.** P(younger)

c. P(boy) **d.** P(older)

NAME ______________________ DATE ____________ PERIOD ____________

12-4 Practice

Probability and Probability Distributions

1. **BALLOONS** A bag contains 1 green, 4 red, and 5 yellow balloons. Two balloons are selected at random. Find the probability of each selection.

   a. P(2 red) **b.** P(1 red and 1 yellow) **c.** P(1 green and 1 yellow)

   d. P(2 green) **e.** P(2 red and 1 yellow) **f.** P(1 red and 1 green)

2. **COINS** A bank contains 3 pennies, 8 nickels, 4 dimes, and 10 quarters. Two coins are selected at random. Find the probability of each selection.

   a. P(2 pennies) **b.** P(2 dimes) **c** P(1 nickel and 1 dime)

   d. P(1 quarter and 1 penny) **e.** P(1 quarter and 1 nickel) **f.** P(2 dimes and 1 quarter)

3. **WALLPAPER** Henrico visits a home decorating store to choose wallpapers for his new house. The store has 28 books of wallpaper samples, including 10 books of WallPride samples and 18 books of Deluxe Wall Coverings samples. The store will allow Henrico to bring 4 books home for a few days so he can decide which wallpapers he wants to buy. If Henrico randomly chooses 4 books to bring home, find the probability of each selection.

   a. P(4 WallPride) **b.** P(2 WallPride and 2 Deluxe)

   c. P(1 WallPride and 3 Deluxe) **d.** P(3 WallPride and 1 Deluxe)

4. **SAT SCORES** The table to the right shows the range of verbal SAT scores for freshmen at a small liberal arts college. If a freshman student is chosen at random, find each probability. Express as decimals rounded to the nearest thousandth.

Range	400–449	450–499	500–549	550–559	600–649	650+
Number of Students	129	275	438	602	620	412

   a. P(400–449) **b.** P(550–559) **c.** P(at least 650)

20. **CHECKERS** The following table shows the wins and losses of the checkers team. If a game is chosen at random, find each probability.

	Arthur	Lynn	Pedro	Mei-Mei
Wins	15	7	12	18
Losses	5	13	3	2

   a. a game was lost and Arthur was playing

   b. Mei-Mei was playing and the game was won

   c. a game was won and Lynn or Arthur was playing

   d. Pedro or Mei-Mei was playing and the game was lost

   e. any of the players was playing and the game was won

NAME ______________________ DATE ____________ PERIOD ____________

12-5 Skills Practice

The Normal Distribution

Determine whether the data appear to be *positively skewed*, *negatively skewed*, or *normally distributed.*

1.

Miles Run	Track Team Members
0–4	3
5–9	4
10–14	7
15–19	5
20–23	2

2.

Speeches Given	Political Candidates
0–5	1
6–11	2
12–17	3
18–23	8
24–29	8

3. PATIENTS The frequency table to the right shows the average number of days patients spent on the surgical ward of a hospital last year.

Days	Number of Patients
0–3	5
4–7	18
8–11	11
12–15	9
16+	6

a. What percentage of the patients stayed between 4 and 7 days?

b. Does the data appear to be *positively skewed*, *negatively skewed*, or *normally distributed*? Explain.

4. DELIVERY The time it takes a bicycle courier to deliver a parcel to his farthest customer is normally distributed with a mean of 40 minutes and a standard deviation of 4 minutes.

a. About what percent of the courier's trips to this customer take between 36 and 44 minutes?

b. About what percent of the courier's trips to this customer take between 40 and 48 minutes?

c. About what percent of the courier's trips to this customer take less than 32 minutes?

5. TESTING The average time it takes sophomores to complete a math test is normally distributed with a mean of 63.3 minutes and a standard deviation of 12.3 minutes.

a. About what percent of the sophomores take more than 75.6 minutes to complete the test?

b. About what percent of the sophomores take between 51 and 63.3 minutes?

c. About what percent of the sophomores take less than 63.3 minutes to complete the test?

NAME _______________ DATE _______ PERIOD _______

12-5 Practice

The Normal Distribution

Determine whether the data appear to be *positively skewed, negatively skewed,* or *normally distributed.*

1.

Time Spent at a Museum Exhibit	
Minutes	**Frequency**
0–25	27
26–50	46
51–75	89
75–100	57
100+	24

2.

Average Age of High School Principals	
Age in Years	**Number**
31–35	3
36–40	8
41–45	15
46–50	32
51–55	40
56–60	38
60+	4

3. STUDENTS The frequency table to the right shows the number of hours worked per week by 100 high school students.

Hours	Number of Students
0–8	30
9–17	45
18–25	20
26+	5

a. What percentage of the students worked between 9 and 17 days?

b. Graph the data. Do the data appear to be *positively skewed, negatively skewed,* or *normally distributed*? Explain.

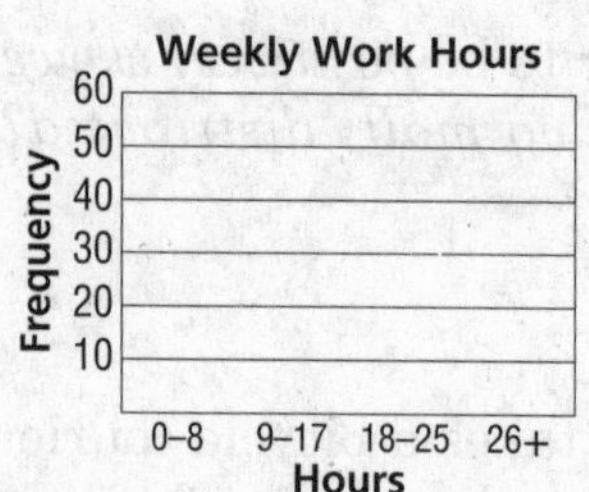

4. TESTING The scores on a test administered to prospective employees are normally distributed with a mean of 100 and a standard deviation of 15.

a. About what percent of the scores are between 70 and 130?

b. About what percent of the scores are between 85 and 130?

c. About what percent of the scores are over 115?

d. About what percent of the scores are lower than 85 or higher than 115?

e. If 80 people take the test, how many would you expect to score higher than 130?

f. If 75 people take the test, how many would you expect to score lower than 85?

5. TEMPERATURE The daily July surface temperature of a lake at a resort has a mean of 82° and a standard deviation of 4.2°. If you prefer to swim when the temperature is at least 77.8°, about what percent of the days does the temperature meet your preference?

NAME ______________________ DATE ____________ PERIOD ________

12-6 Skills Practice

Hypothesis Testing

Find a 95% confidence interval for each of the following. Round to the nearest tenth, if necessary.

1. $\bar{x} = 21$, $s = 3$, and $n = 10{,}000$

2. $\bar{x} = 50$, $s = 2.5$, and $n = 50$

3. $\bar{x} = 120$, $s = 9$, and $n = 144$

4. $\bar{x} = 10.5$, $s = 7.9$, and $n = 100$

5. $\bar{x} = 200$, $s = 18$, and $n = 120$

6. $\bar{x} = 21$, $s = 4$, and $n = 50$

7. $\bar{x} = 58$, $s = 3.5$, and $n = 7$

8. $\bar{x} = 84$, $s = 5$, and $n = 100$

9. $\bar{x} = 115.1$, $s = 12.8$, and $n = 200$

10. $\bar{x} = 48$, $s = 7.5$, and $n = 150$

Test each null hypothesis. Write *accept* or *reject*.

11. $H_0 = 72$, $H_1 < 72$, $n = 50$, $\bar{x} = 71.3$, and $\sigma = 2$

12. $H_0 = 45$, $H_1 < 45$, $n = 50$, $\bar{x} = 40$, and $\sigma = 7$

13. $H_0 = 11.7$, $H_1 > 11.7$, $n = 100$, $\bar{x} = 12$ and $\sigma = 1.5$

14. $H_0 = 151.3$, $H_1 < 151.3$, $n = 150$, $\bar{x} = 150$, and $\sigma = 4$

15. $H_0 = 40$, $H_1 > 40$, $n = 5$, $\bar{x} = 42$, and $\sigma = 2$

16. $H_0 = 100.5$, $H_1 < 100.5$, $n = 256$, $\bar{x} = 100$, and $\sigma = 4$

17. $H_0 = 26$, $H_1 > 26$, $n = 2000$, $\bar{x} = 28$, and $\sigma = 4.5$

18. $H_0 = 68.7$, $H_1 > 68.7$, $n = 196$, $\bar{x} = 70.7$, and $\sigma = 14$

19. $H_0 = 7$, $H_1 > 7$, $n = 100$, $\bar{x} = 7.2$, and $\sigma = 1$

20. $H_0 = 55.63$, $H_1 < 55.63$, $n = 100$, $\bar{x} = 55$, and $\sigma = 3.2$

12-6 Practice

Hypothesis Testing

Find a 95% confidence interval for each of the following. Round to the nearest tenth, if necessary.

1. $\bar{x} = 56$, $s = 2$, and $n = 50$

2. $\bar{x} = 99$, $s = 22$, and $n = 121$

3. $\bar{x} = 34$, $s = 4$, and $n = 200$

4. $\bar{x} = 12$, $s = 4.5$, and $n = 100$

5. $\bar{x} = 37$, $s = 2.5$, and $n = 50$

6. $\bar{x} = 78$, $s = 2$, and $n = 225$

7. $\bar{x} = 36$, $s = 6$, and $n = 36$

8. $\bar{x} = 121$, $s = 2.5$, and $n = 100$

Test each null hypothesis. Write *accept* or *reject*.

9. $H_0 = 200.1$, $H_1 < 200.1$, $n = 200$, $\bar{x} = 50$, and $\sigma = 2$

10. $H_0 = 75.6$, $H_1 < 75.6$, $n = 100$, $\bar{x} = 77$, and $\sigma = 7$

11. $H_0 = 89.3$, $H_1 < 89.3$, $n = 100$, $\bar{x} = 89$ and $\sigma = 1.5$

12 $H_0 = 75$, $H_1 < 75$, $n = 150$, $\bar{x} = 74.2$, and $\sigma = 2.5$

13. $H_0 = 121$, $H_1 < 121$, $n = 64$, $\bar{x} = 120$, and $\sigma = 2$

14. $H_0 = 198.5$, $H_1 > 198.5$, $n = 100$, $\bar{x} = 200$, and $\sigma = 7.5$

15. $H_0 = 38.5$, $H_1 > 38.5$, $n = 50$, $\bar{x} = 40$, and $\sigma = 4.5$

16. $H_0 = 112.5$, $H_1 < 112.5$, $n = 100$, $\bar{x} = 110.5$, and $\sigma = 10$

17. RUNNING Josh and his sister Megan run together each morning and do not use a stopwatch to keep track of their time. Josh thinks they usually run the mile under 7 minutes, while Megan thinks it takes them longer. They borrow a stopwatch and time themselves each day for 20 days. Their mean time to run one mile is 7.4 minutes with a standard deviation of 0.2 minutes. Test Megan's hypothesis.

18. QUALITY CONTROL Kim is a quality tester for a tropical fruit company. The company claims that their canned pineapple stays fresh for at least 16 hours after opening. Kim tests 15 different cans to see if they actually stay fresh for at least 16 hours. Use the data below to conduct a hypothesis test.

Number of Hours Each Can Stays Fresh				
12	14	7	12	10
12	12	13	16	9
5	11	19	18	6

NAME ______________________ DATE ____________ PERIOD ____________

12-7 Skills Practice

Binomial Distributions

1. COINS Find each probability if a coin is tossed 4 times.

a. P(4 heads)

b. P(0 heads)

c. P(exactly 3 heads)

d. P(exactly 2 heads)

e. P(exactly 1 head)

f. P(at least 3 heads)

2. DICE Find each probability if a die is rolled 3 times.

a. P(exactly one 2)

b. P(exactly two 2s)

c. P(exactly three 2s)

d. P(at most one 2)

3. FIREWORKS A town that presents a fireworks display during its July 4 celebration found the probability that a family with two or more children will watch the fireworks is $\frac{3}{5}$. If 5 of these families are selected at random, find each probability.

a. P(exactly 3 families watch the fireworks)

b. P(exactly 2 families watch the fireworks)

c. P(exactly 5 families watch the fireworks)

d. P(no families watch the fireworks)

e. P(at least 4 families watch the fireworks)

f. P(at most 1 family watches the fireworks)

4. TESTS One section of a standardized English language test has 10 true/false questions. Find each probability when a student guesses at all ten questions.

a. P(exactly 8 correct)

b. P(exactly 2 correct)

c. P(exactly half correct)

d. P(all 10 correct)

e. P(0 correct)

f. P(at least 8 correct)

NAME ______________________ DATE ____________ PERIOD ____________

12-7 Practice

Binomial Distributions

1. COINS Find each probability if a coin is tossed 6 times.

a. *P*(exactly 3 tails)

b. *P*(exactly 5 tails)

c. *P*(0 tails)

d. *P*(at least 4 heads)

e. *P*(at least 4 tails)

f. *P*(at most 2 tails)

2. FREE THROWS The probability of Chris making a free throw is $\frac{2}{3}$. If she shoots 5 times, find each probability.

a. *P*(all missed)

b. *P*(all made)

c. *P*(exactly 2 made)

d. *P*(exactly 1 missed)

e. *P*(at least 3 made)

f. *P*(at most 2 made)

3. BOARD GAME When Tarin and Sam play a certain board game, the probability that Tarin will win a game is $\frac{3}{4}$. If they play 5 games, find each probability.

a. *P*(Sam wins only once)

b. *P*(Tarin wins exactly twice)

c. *P*(Sam wins exactly 3 games)

d. *P*(Sam wins at least 1 game)

e. *P*(Tarin wins at least 3 games)

f. *P*(Tarin wins at most 2 games)

4. SAFETY In August 2001, the American Automobile Association reported that 73% of Americans use seat belts. In a random selection of 10 Americans in 2001, what is the probability that exactly half of them use seat belts?

5. HEALTH In 2001, the American Heart Association reported that 50 percent of the Americans who receive heart transplants are ages 50–64 and 20 percent are ages 35–49.

a. In a randomly selected group of 10 heart transplant recipients, what is the probability that at least 8 of them are ages 50–64?

b. In a randomly selected group of 5 heart transplant recipients, what is the probability that 2 of them are ages 35–49?

NAME ______________________ DATE ____________ PERIOD ____________

13-1 Skills Practice

Trigonometric Functions in Right Triangles

Find the values of the six trigonometric functions for angle θ.

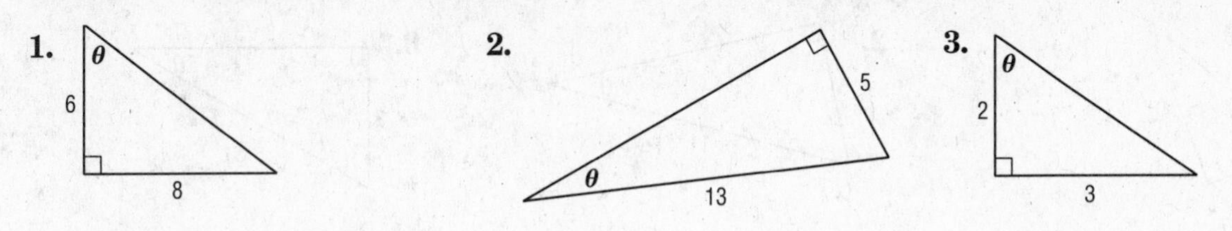

In a right triangle, $\angle A$ is acute.

4. If $\tan A = 3$, what is $\sin A$?

5. If $\sin A = \frac{1}{16}$, what is $\cos A$?

Use a trigonometric function to find the value of x. Round to the nearest tenth if necessary.

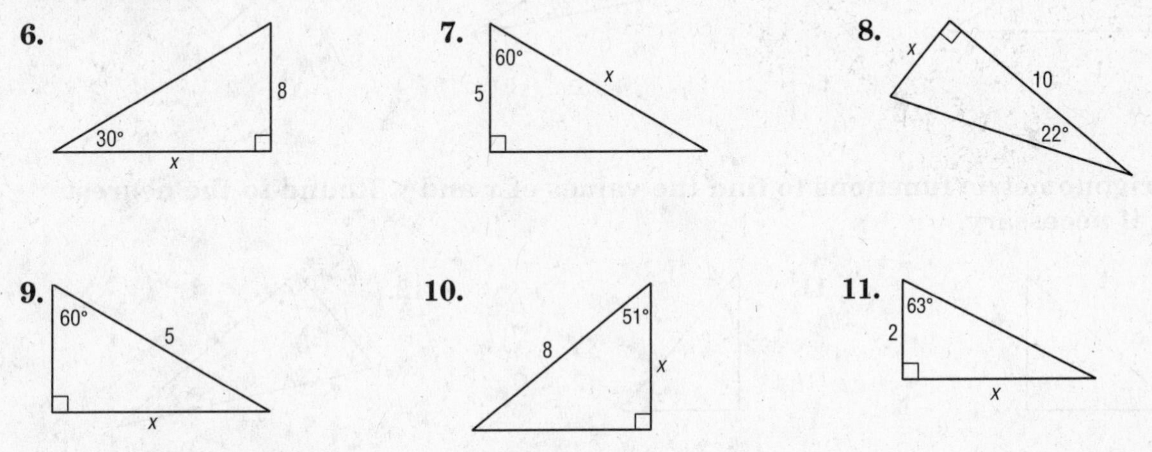

Find the value of x. Round to the nearest tenth if necessary.

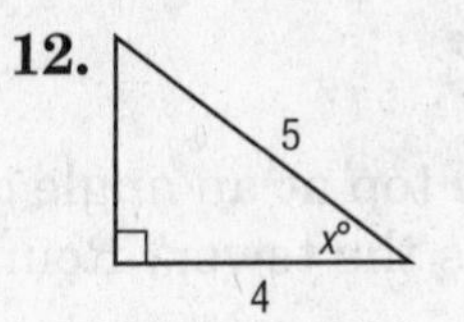

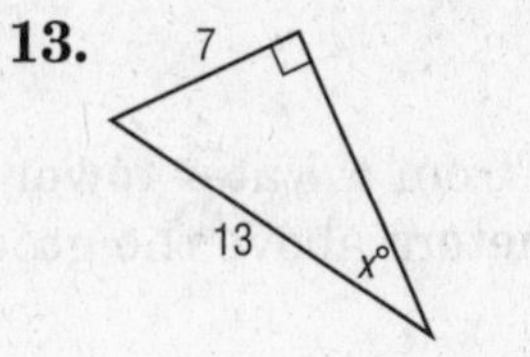

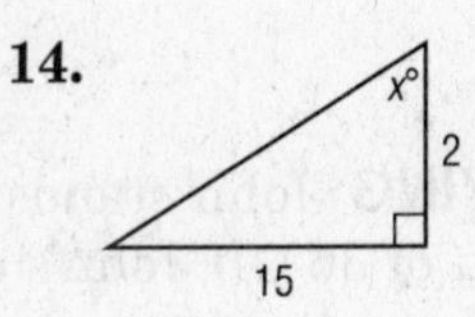

13-1 Practice

Trigonometric Functions in Right Triangles

Find the values of the six trigonometric functions for angle θ.

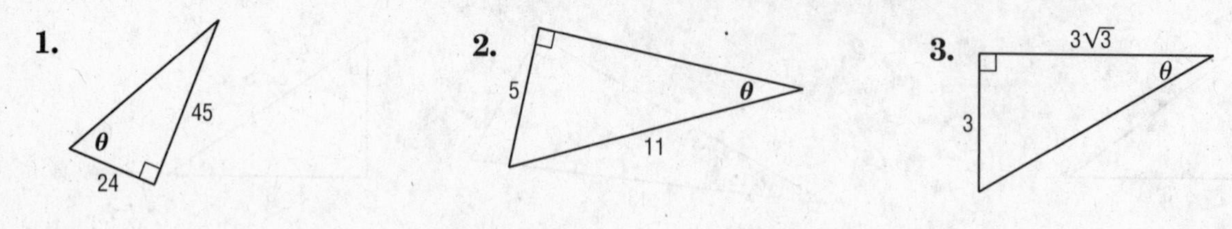

In a right triangle, $\angle A$ and $\angle B$ are acute.

4. If $\tan B = 2$, what is $\cos B$? **5.** If $\tan A = \frac{11}{17}$, what is $\sin A$? **6.** If $\sin B = \frac{8}{15}$, what is $\cos B$?

Use a trigonometric function to find each value of x. Round to the nearest tenth if necessary.

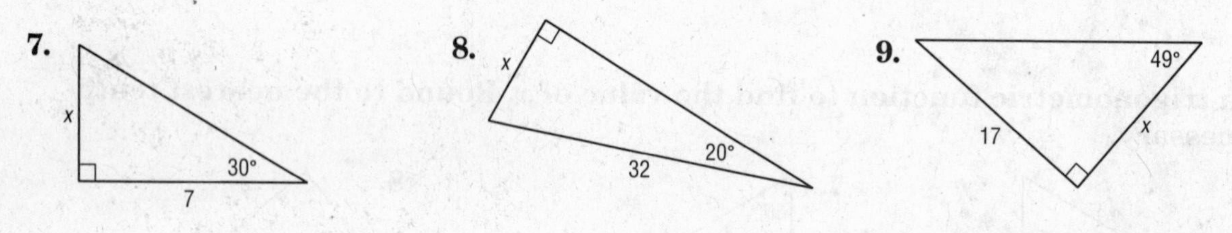

Use trigonometric functions to find the values of x and y. Round to the nearest tenth if necessary.

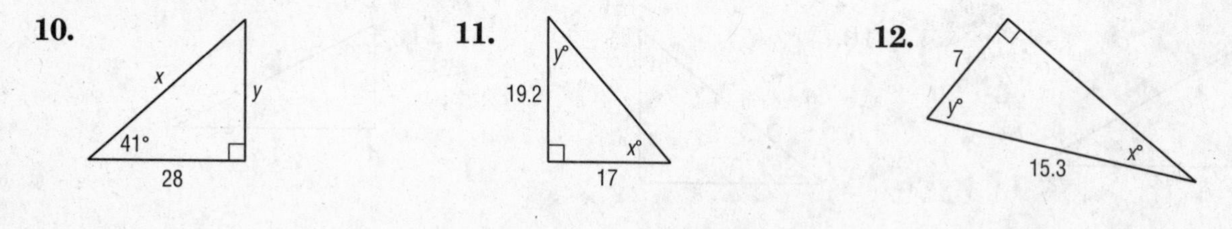

13. SURVEYING John stands 150 meters from a water tower and sights the top at an angle of elevation of 36°. If John's eyes are 2 meters above the ground, how tall is the tower? Round to the nearest meter.

NAME ______________________ DATE ____________ PERIOD ____________

13-2 Skills Practice

Angles and Angle Measure

Draw an angle with the given measure in standard position.

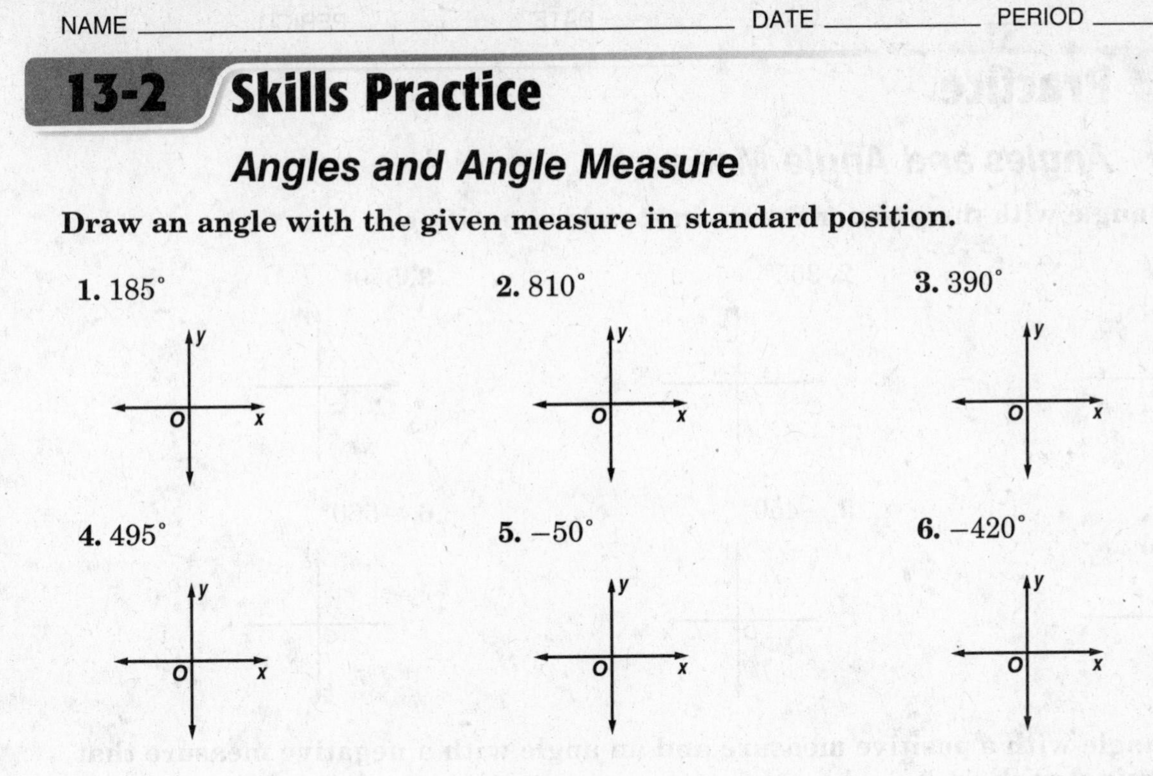

Find an angle with a positive measure and an angle with a negative measure that are coterminal with each angle.

7. 45°

8. 60°

9. 370°

10. –90°

11. $\frac{2\pi}{3}$

12. $\frac{5\pi}{2}$

13. $\frac{\pi}{6}$

14. $-\frac{3\pi}{4}$

Rewrite each degree measure in radians and each radian measure in degrees.

15. 130°

16. 720°

17. 210°

18. 90°

19. –30°

20. –270°

21. $\frac{\pi}{3}$

22. $\frac{5\pi}{6}$

23. $\frac{2\pi}{3}$

24. $\frac{5\pi}{4}$

25. $-\frac{3\pi}{4}$

26. $-\frac{7\pi}{6}$

NAME ______________________ DATE ____________ PERIOD ____________

13-2 Practice

Angles and Angle Measure

Draw an angle with the given measure in standard position.

1. 210° **2.** 305° **3.** 580°

4. 135° **5.** −450° **6.** −560°

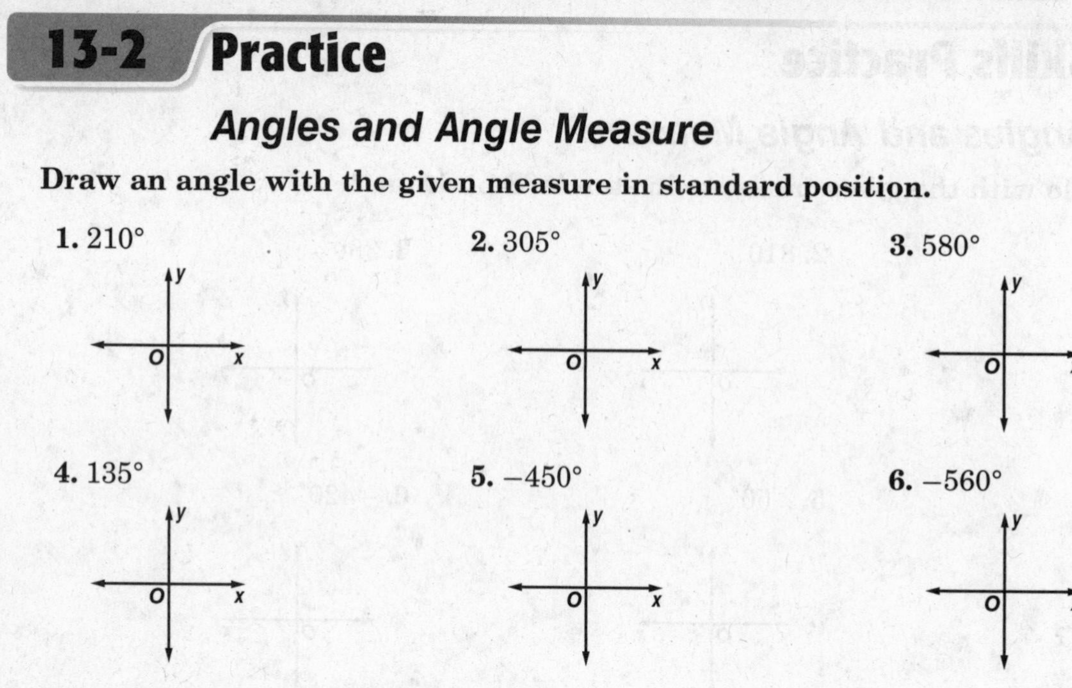

Find an angle with a positive measure and an angle with a negative measure that are coterminal with each angle.

7. 65° **8.** 80° **9.** 110°

10. $\frac{2\pi}{5}$ **11.** $\frac{5\pi}{6}$ **12.** $-\frac{3\pi}{2}$

Rewrite each degree measure in radians and each radian measure in degrees.

13. 18° **14.** 6° **15.** −72° **16.** −820°

17. 4π **18.** $\frac{5\pi}{2}$ **19.** $-\frac{9\pi}{2}$ **20.** $-\frac{7\pi}{12}$

Find the length of each arc. Round to the nearest tenth.

21. **22.** **23.**

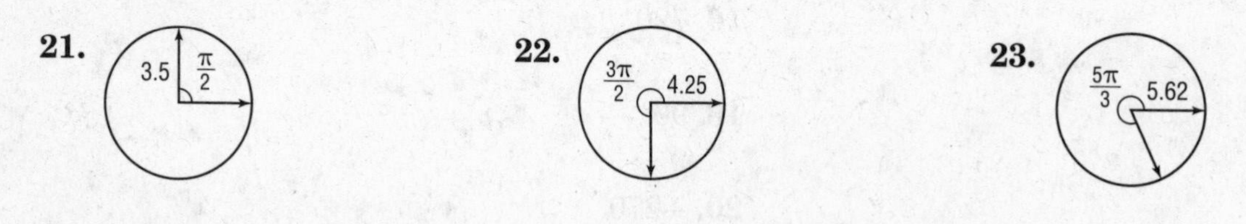

24. TIME Find both the degree and radian measures of the angle through which the hour hand on a clock rotates from 5 A.M. to 10 P.M.

25. ROTATION A truck with 16-inch radius wheels is driven at 77 feet per second (52.5 miles per hour). Find the measure of the angle through which a point on the outside of the wheel travels each second. Round to the nearest degree and nearest radian.

NAME ______________________ DATE __________ PERIOD __________

13-3 Skills Practice

Trigonometric Functions of General Angles

The terminal side of θ in standard position contains each point. Find the exact values of the six trigonometric functions of θ.

1. (5, 12)

2. (3, 4)

3. (8, −15)

4. (−4, 3)

5. (−9, −40)

6. (1, 2)

7. (3, −9)

8. (−8, 12)

Sketch each angle. Then find its reference angle.

9. 135°

10. 200°

11. $\frac{5\pi}{3}$

Find the exact value of each trigonometric function.

12. sin 150°

13. cos 270°

14. cot 135°

15. tan (−30°)

16. $\tan \frac{\pi}{4}$

17. $\cos \frac{4\pi}{3}$

18. cot (−π)

19. $\sin\left(-\frac{3\pi}{4}\right)$

NAME ______________________________ DATE ____________ PERIOD ____________

13-3 Practice

Trigonometric Functions of General Angles

The terminal side of θ in standard position contains each point. Find the exact values of the six trigonometric functions of θ.

1. (6, 8) **2.** (−20, 21) **3.** (−2, −5)

Sketch each angle. Then find its reference angle.

4. $\frac{13\pi}{8}$ **5.** −210° **6.** $-\frac{7\pi}{4}$

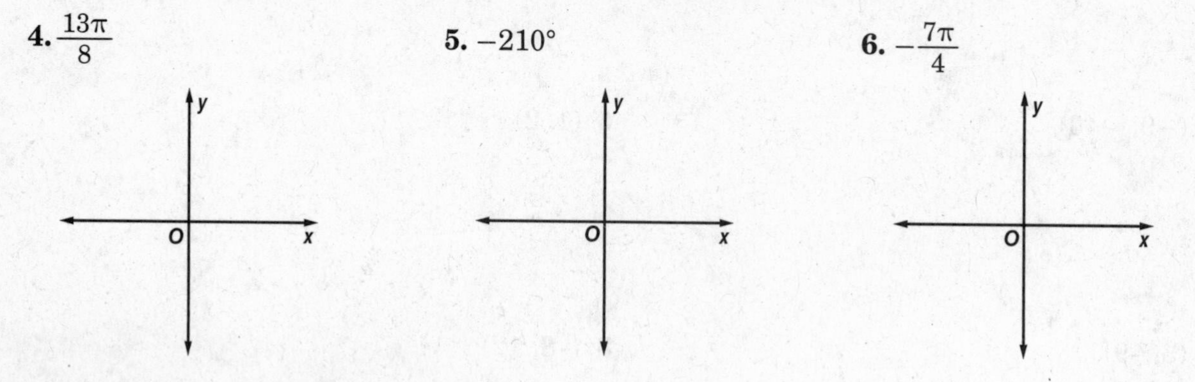

Find the exact value of each trigonometric function.

7. tan 135° **8.** cot 210° **9.** cot (−90°) **10.** cos 405°

11. $\tan \frac{5\pi}{3}$ **12.** $\csc \left(-\frac{3\pi}{4}\right)$ **13.** cot 2π **14.** $\tan \frac{13\pi}{6}$

15. LIGHT Light rays that "bounce off" a surface are *reflected* by the surface. If the surface is partially transparent, some of the light rays are bent or *refracted* as they pass from the air through the material. The angles of reflection θ_1 and of refraction θ_2 in the diagram at the right are related by the equation $\sin \theta_1 = n \sin \theta_2$. If $\theta_1 = 60°$ and $n = \sqrt{3}$, find the measure of θ_2.

16. FORCE A cable running from the top of a utility pole to the ground exerts a horizontal pull of 800 Newtons and a vertical pull of $800\sqrt{3}$ Newtons. What is the sine of the angle θ between the cable and the ground? What is the measure of this angle?

NAME ______________________ DATE ____________ PERIOD ________

13-4 Skills Practice

Law of Sines

Find the area of $\triangle ABC$ to the nearest tenth, if necessary.

1.

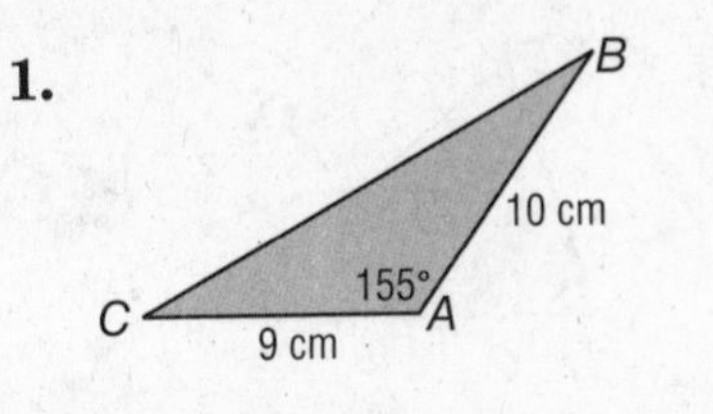

2.

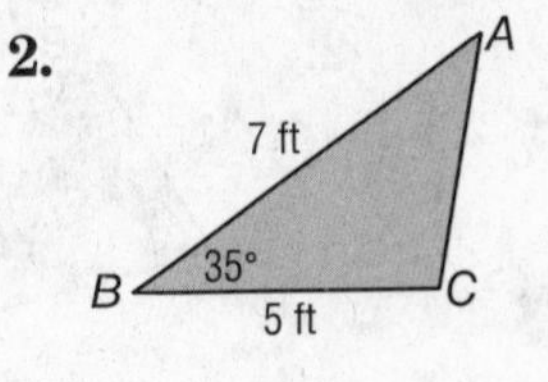

3. $A = 35°$, $b = 3$ ft, $c = 7$ ft

4. $C = 148°$, $a = 10$ cm, $b = 7$ cm

5. $C = 22°$, $a = 14$ m, $b = 8$ m

6. $B = 93°$, $c = 18$ mi, $a = 42$ mi

Solve each triangle. Round side lengths to the nearest tenth and angle measures to the nearest degree.

7. **8.** **9.**

10. **11.** **12.**

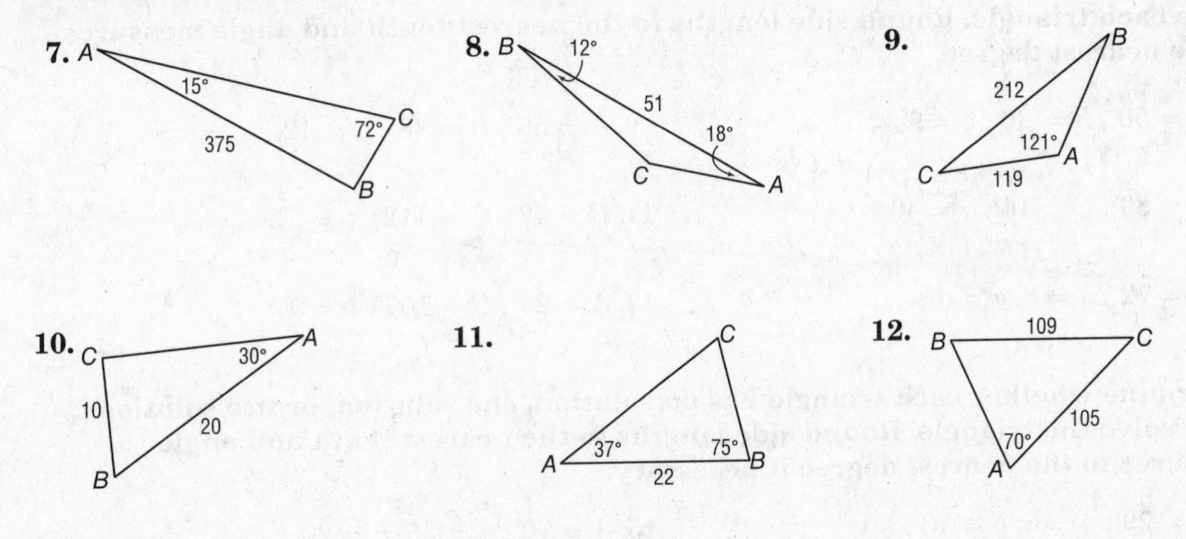

Determine whether each triangle has *no* solution, *one* solution, or *two* solutions. Then solve the triangle. Round side lengths to the nearest tenth and angle measures to the nearest degree.

13. $A = 30°$, $a = 1$, $b = 4$

14. $A = 30°$, $a = 2$, $b = 4$

15. $A = 30°$, $a = 3$, $b = 4$

16. $A = 38°$, $a = 10$, $b = 9$

17. $A = 78°$, $a = 8$, $b = 5$

18. $A = 133°$, $a = 9$, $b = 7$

19. $A = 127°$, $a = 2$, $b = 6$

20. $A = 109°$, $a = 24$, $b = 13$

NAME ______________________ DATE ____________ PERIOD ________

13-4 Practice

Law of Sines

Find the area of $\triangle ABC$ to the nearest tenth, if necessary.

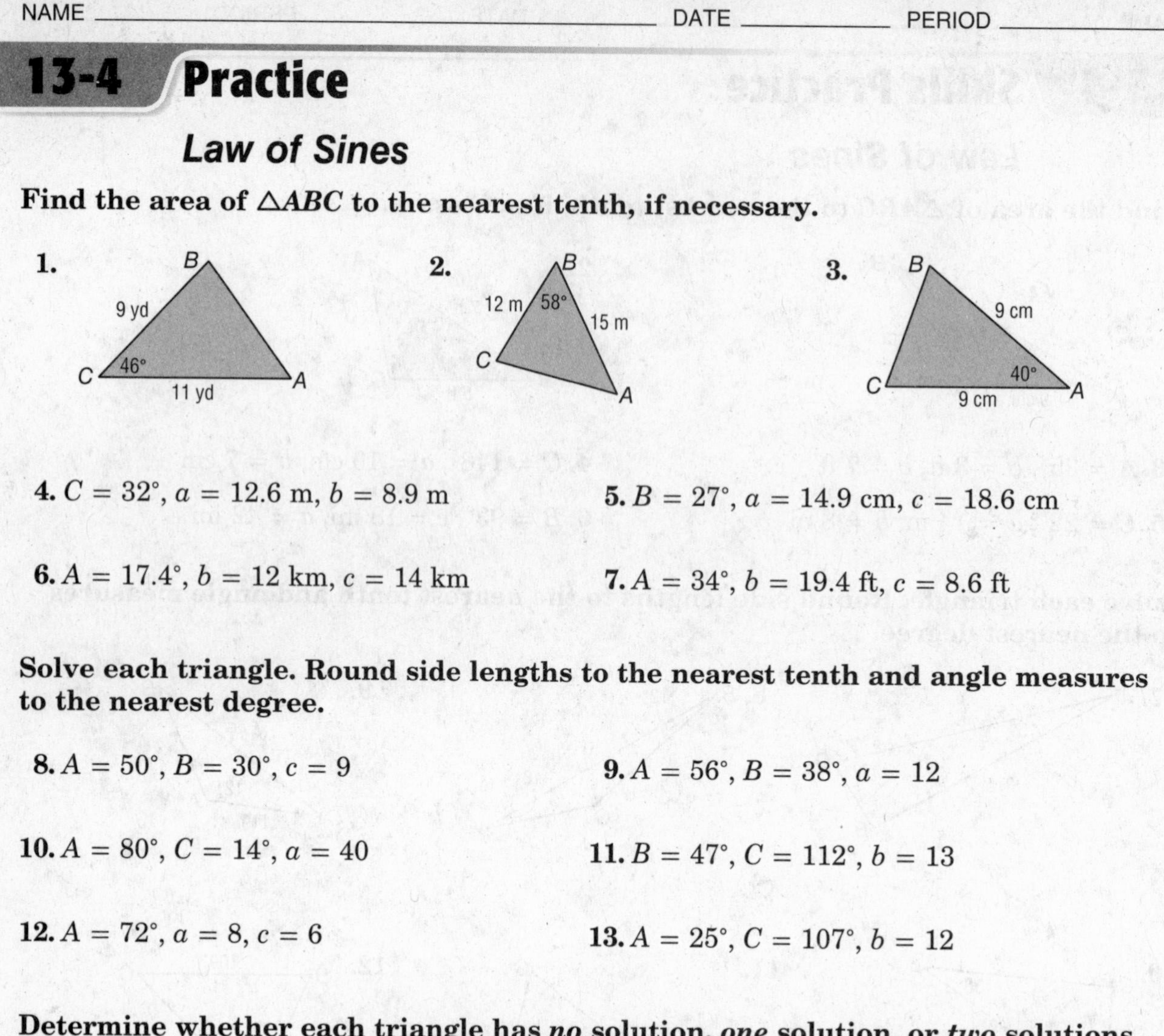

4. $C = 32°$, $a = 12.6$ m, $b = 8.9$ m

5. $B = 27°$, $a = 14.9$ cm, $c = 18.6$ cm

6. $A = 17.4°$, $b = 12$ km, $c = 14$ km

7. $A = 34°$, $b = 19.4$ ft, $c = 8.6$ ft

Solve each triangle. Round side lengths to the nearest tenth and angle measures to the nearest degree.

8. $A = 50°, B = 30°, c = 9$

9. $A = 56°, B = 38°, a = 12$

10. $A = 80°, C = 14°, a = 40$

11. $B = 47°, C = 112°, b = 13$

12. $A = 72°, a = 8, c = 6$

13. $A = 25°, C = 107°, b = 12$

Determine whether each triangle has *no* solution, *one* solution, or *two* solutions. Then solve the triangle. Round side lengths to the nearest tenth and angle measures to the nearest degree if necessary.

14. $A = 29°, a = 6, b = 13$

15. $A = 70°, a = 25, b = 20$

16. $A = 113°, a = 21, b = 25$

17. $A = 110°, a = 20, b = 8$

18. $A = 66°, a = 12, b = 7$

19. $A = 54°, a = 5, b = 8$

20. $A = 45°, a = 15, b = 18$

21. $A = 60°, a = 4\sqrt{3}, b = 8$

22. WILDLIFE Sarah Phillips, an officer for the Department of Fisheries and Wildlife, checks boaters on a lake to make sure they do not disturb two osprey nesting sites. She leaves a dock and heads due north in her boat to the first nesting site. From here, she turns 5° north of due west and travels an additional 2.14 miles to the second nesting site. She then travels 6.7 miles directly back to the dock. How far from the dock is the first osprey nesting site? Round to the nearest tenth.

13-5 Skills Practice

Law of Cosines

Solve each triangle. Round side lengths to the nearest tenth and angle measures to the nearest degree.

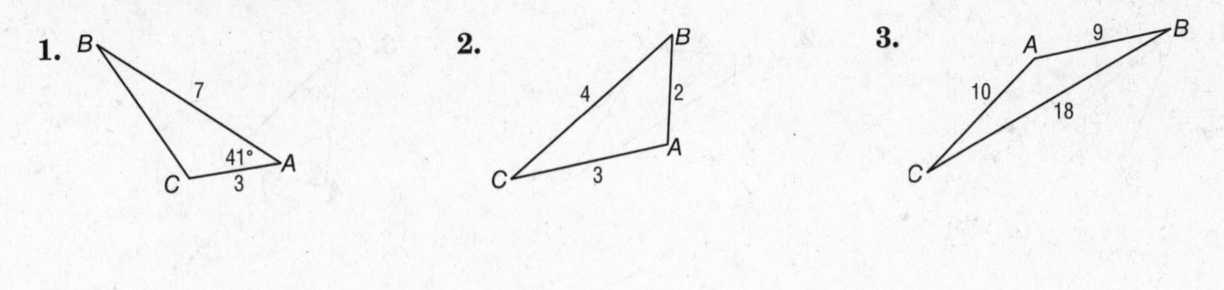

4. $C = 71^\circ$, $a = 3$, $b = 4$

5. $C = 35^\circ$, $a = 5$, $b = 8$

Determine whether each triangle should be solved by begining with the Law of *Sines* or the Law of *Cosines*. Then solve the triangle.

6.

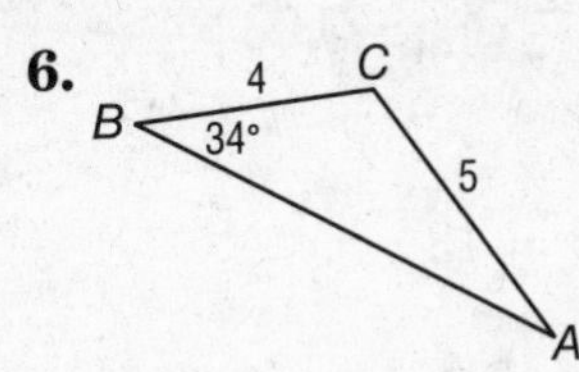

7.

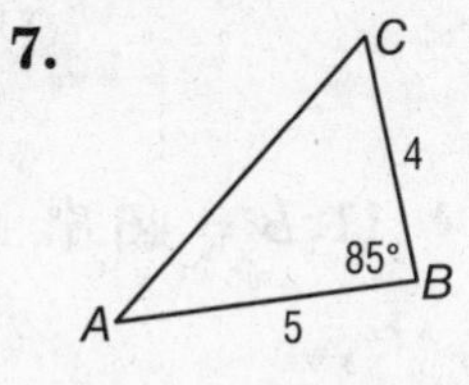

8.

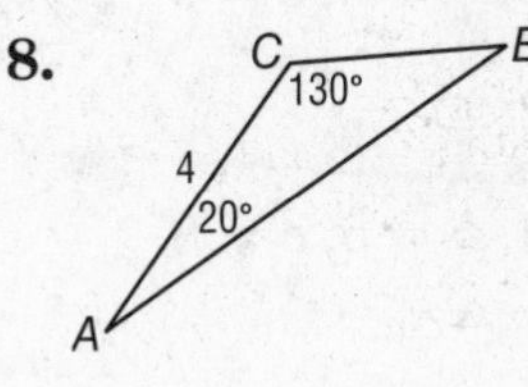

9. $A = 11^\circ$, $C = 27^\circ$, $c = 50$

10. $B = 47^\circ$, $a = 20$, $c = 24$

11. $A = 71^\circ$, $C = 62^\circ$, $a = 20$

12. $a = 5$, $b = 12$, $c = 13$

13. $A = 51^\circ$, $b = 7$, $c = 10$

14. $a = 13$, $A = 41^\circ$, $B = 75^\circ$

15. $B = 125^\circ$, $a = 8$, $b = 14$

16. $a = 5$, $b = 6$, $c = 7$

NAME ______________________ DATE ____________ PERIOD ________

13-5 Practice

Law of Cosines

Determine whether each triangle should be solved by beginning with the Law of *Sines* or Law of *Cosines*. Then solve the triangle.

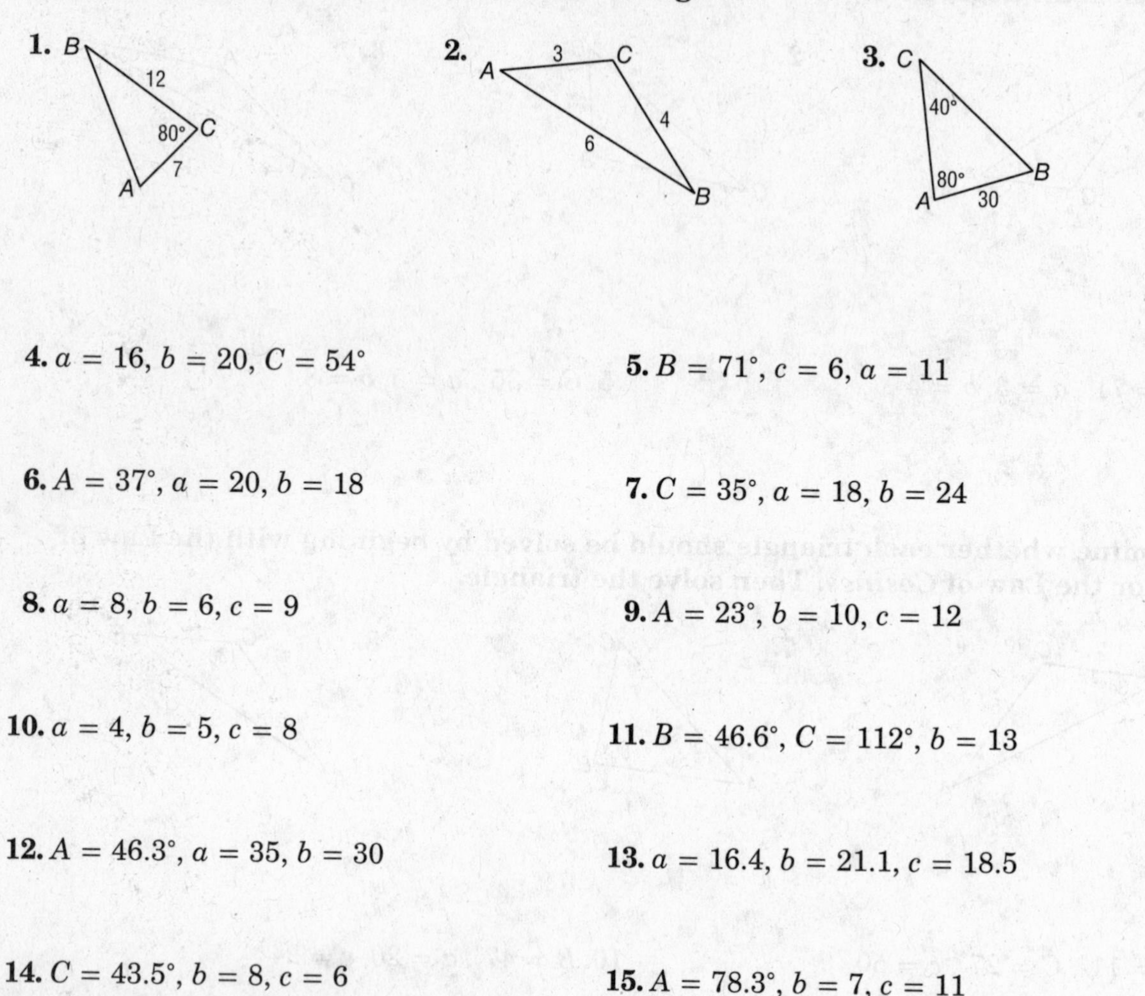

4. $a = 16$, $b = 20$, $C = 54°$

5. $B = 71°$, $c = 6$, $a = 11$

6. $A = 37°$, $a = 20$, $b = 18$

7. $C = 35°$, $a = 18$, $b = 24$

8. $a = 8$, $b = 6$, $c = 9$

9. $A = 23°$, $b = 10$, $c = 12$

10. $a = 4$, $b = 5$, $c = 8$

11. $B = 46.6°$, $C = 112°$, $b = 13$

12. $A = 46.3°$, $a = 35$, $b = 30$

13. $a = 16.4$, $b = 21.1$, $c = 18.5$

14. $C = 43.5°$, $b = 8$, $c = 6$

15. $A = 78.3°$, $b = 7$, $c = 11$

16. SATELLITES Two radar stations 2.4 miles apart are tracking an airplane. The straight-line distance between Station A and the plane is 7.4 miles. The straight-line distance between Station B and the plane is 6.9 miles. What is the angle of elevation from Station A to the plane? Round to the nearest degree.

7.4 mi
6.9 mi
A
B
2.4 mi

17. DRAFTING Marion is using a computer-aided drafting program to produce a drawing for a client. She begins a triangle by drawing a segment 4.2 inches long from point A to point B. From B, she draws a second segment that forms a 42° angle with $\overline{AB}$ and is 6.4 inches long, ending at point C. To the nearest tenth, how long is the segment from C to A?

NAME ______________________ DATE ____________ PERIOD ________

13-6 Skill Practice

Circular Functions

The terminal side of angle θ in standard position intersects the unit circle at each point P. Find $\cos\theta$ and $\sin\theta$.

1. $P\left(\frac{3}{5}, \frac{4}{5}\right)$

2. $P\left(\frac{5}{13}, -\frac{12}{13}\right)$

3. $P\left(-\frac{9}{41}, -\frac{40}{41}\right)$

4. $P(0, 1)$

5. $P(-1, 0)$

6. $P\left(\frac{1}{2}, -\frac{\sqrt{3}}{2}\right)$

Find the exact value of each function.

7. $\cos 45°$

8. $\sin 210°$

9. $\sin 330°$

10. $\cos 330°$

11. $\cos(-60°)$

12. $\sin(-390°)$

13. $\sin 5\pi$

14. $\cos 3\pi$

15. $\sin \frac{5\pi}{2}$

16. $\sin \frac{7\pi}{3}$

17. $\cos\left(-\frac{7\pi}{3}\right)$

18. $\cos\left(-\frac{5\pi}{6}\right)$

Determine the period of each function.

19.

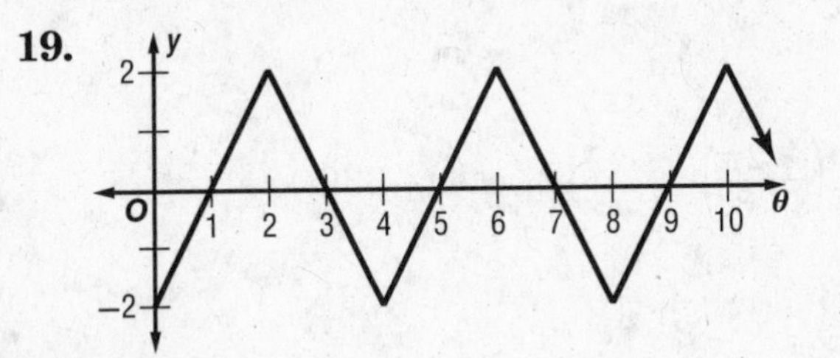

20.

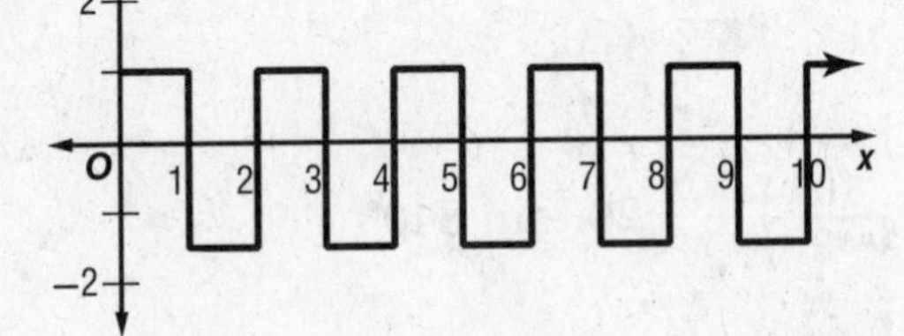

21.

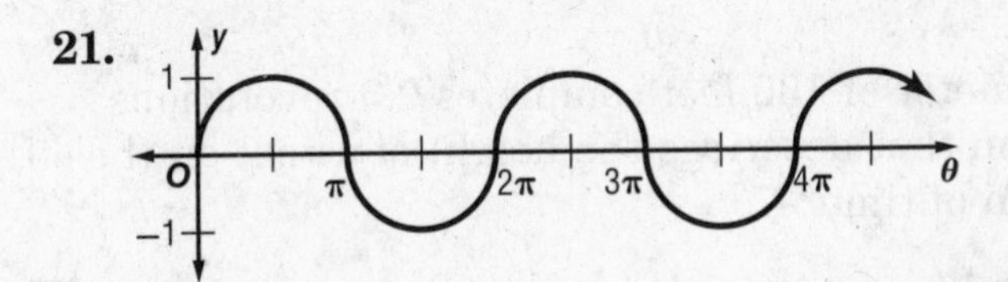

13-6 Practice

Circular Functions

The terminal side of angle θ in standard position intersects the unit circle at each point P. Find $\cos \theta$ and $\sin \theta$.

1. $P\left(-\frac{1}{2}, \frac{\sqrt{3}}{2}\right)$

2. $P\left(\frac{20}{29}, -\frac{21}{29}\right)$

3. $P(0.8, 0.6)$

4. $P(0, -1)$

5. $P\left(-\frac{\sqrt{2}}{2}, -\frac{\sqrt{2}}{2}\right)$

6. $P\left(\frac{\sqrt{3}}{2}, \frac{1}{2}\right)$

Determine the period of each function.

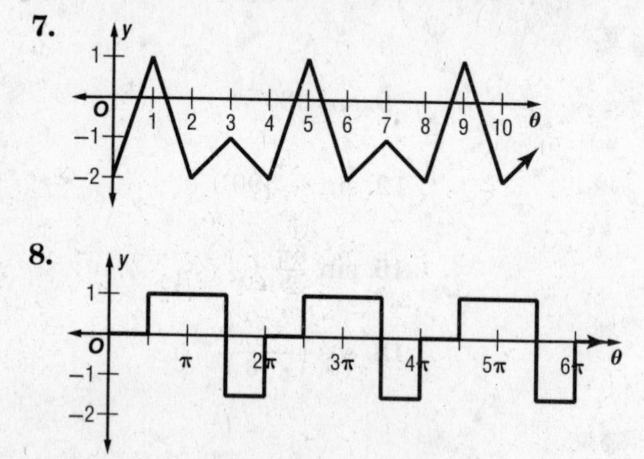

Find the exact value of each function.

9. $\cos \frac{7\pi}{4}$

10. $\sin(-30°)$

11. $\sin\left(-\frac{2\pi}{3}\right)$

12. $\cos(-330°)$

13. $\cos 600°$

14. $\sin \frac{9\pi}{2}$

15. $\cos 7\pi$

16. $\cos\left(-\frac{11\pi}{4}\right)$

17. $\sin(-225°)$

18. $\sin 585°$

19. $\cos\left(-\frac{10\pi}{3}\right)$

20. $\sin 840°$

21. FERRIS WHEELS A Ferris wheel with a diameter of 100 feet completes 2.5 revolutions per minute. What is the period of the function that describes the height of a seat on the outside edge of the Ferris wheel as a function of time?

NAME ______________________ DATE ____________ PERIOD ________

13-7 Skills Practice

Graphing Trigonometric Functions

Find the amplitude and period of each function. Then graph the function.

1. $y = 2 \cos \theta$

2. $y = 4 \sin \theta$

3. $y = 2 \sec \theta$

4. $y = \frac{1}{2} \tan \theta$

5. $y = \sin 3\theta$

6. $y = \csc 3\theta$

7. $y = \tan 2\theta$

8. $y = \cos 2\theta$

9. $y = 4 \sin \frac{1}{2}\theta$

NAME ______________________ DATE ____________ PERIOD ____________

13-7 Practice

Graphing Trigonometric Functions

Find the amplitude, if it exists, and period of each function. Then graph the function.

1. $y = -4 \sin \theta$

2. $y = \cot \frac{1}{2}\theta$

3. $y = \cos 5\theta$

4. $y = \csc \frac{3}{4}\theta$

5. $y = 2 \tan \frac{1}{2}\theta$

6. $y = \frac{1}{2} \sin \theta$

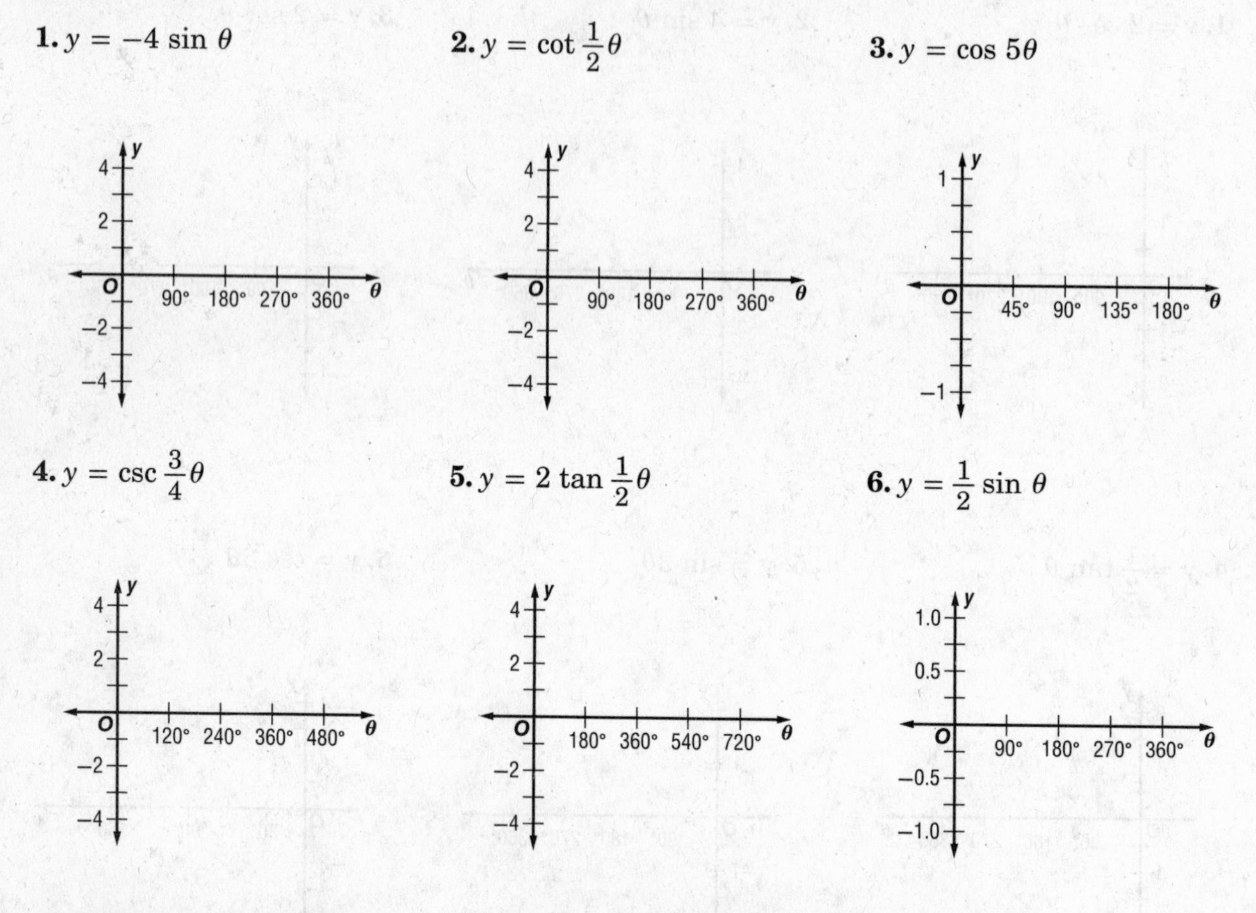

7. FORCE An anchoring cable exerts a force of 500 Newtons on a pole. The force has the horizontal and vertical components F_x and F_y. (A force of one Newton (N), is the force that gives an acceleration of 1 m/sec² to a mass of 1 kg.)

500 N
F_y
θ
F_x

a. The function $F_x = 500 \cos \theta$ describes the relationship between the angle θ and the horizontal force. What are the amplitude and period of this function?

b. The function $F_y = 500 \sin \theta$ describes the relationship between the angle θ and the vertical force. What are the amplitude and period of this function?

8. WEATHER The function $y = 60 + 25 \sin \frac{\pi}{6}t$, where t is in months and $t = 0$ corresponds to April 15, models the average high temperature in degrees Fahrenheit in Centerville.

a. Determine the period of this function. What does this period represent?

b. What is the maximum high temperature and when does this occur?

NAME ______________________ DATE ____________ PERIOD ________

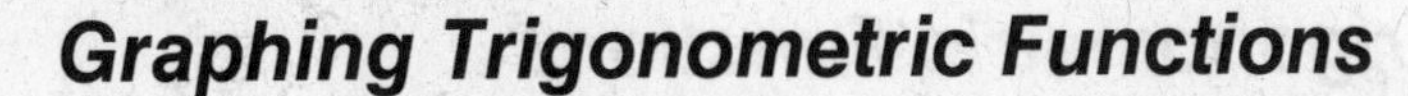

13-7 Skills Practice

Graphing Trigonometric Functions

Find the amplitude and period of each function. Then graph the function.

1. $y = 2 \cos \theta$

2. $y = 4 \sin \theta$

3. $y = 2 \sec \theta$

4. $y = \frac{1}{2} \tan \theta$

5. $y = \sin 3\theta$

6. $y = \csc 3\theta$

7. $y = \tan 2\theta$

8. $y = \cos 2\theta$

9. $y = 4 \sin \frac{1}{2}\theta$

NAME ______________________ DATE ____________ PERIOD ____________

13-7 Practice

Graphing Trigonometric Functions

Find the amplitude, if it exists, and period of each function. Then graph the function.

1. $y = -4 \sin \theta$

2. $y = \cot \frac{1}{2}\theta$

3. $y = \cos 5\theta$

4. $y = \csc \frac{3}{4}\theta$

5. $y = 2 \tan \frac{1}{2}\theta$

6. $y = \frac{1}{2} \sin \theta$

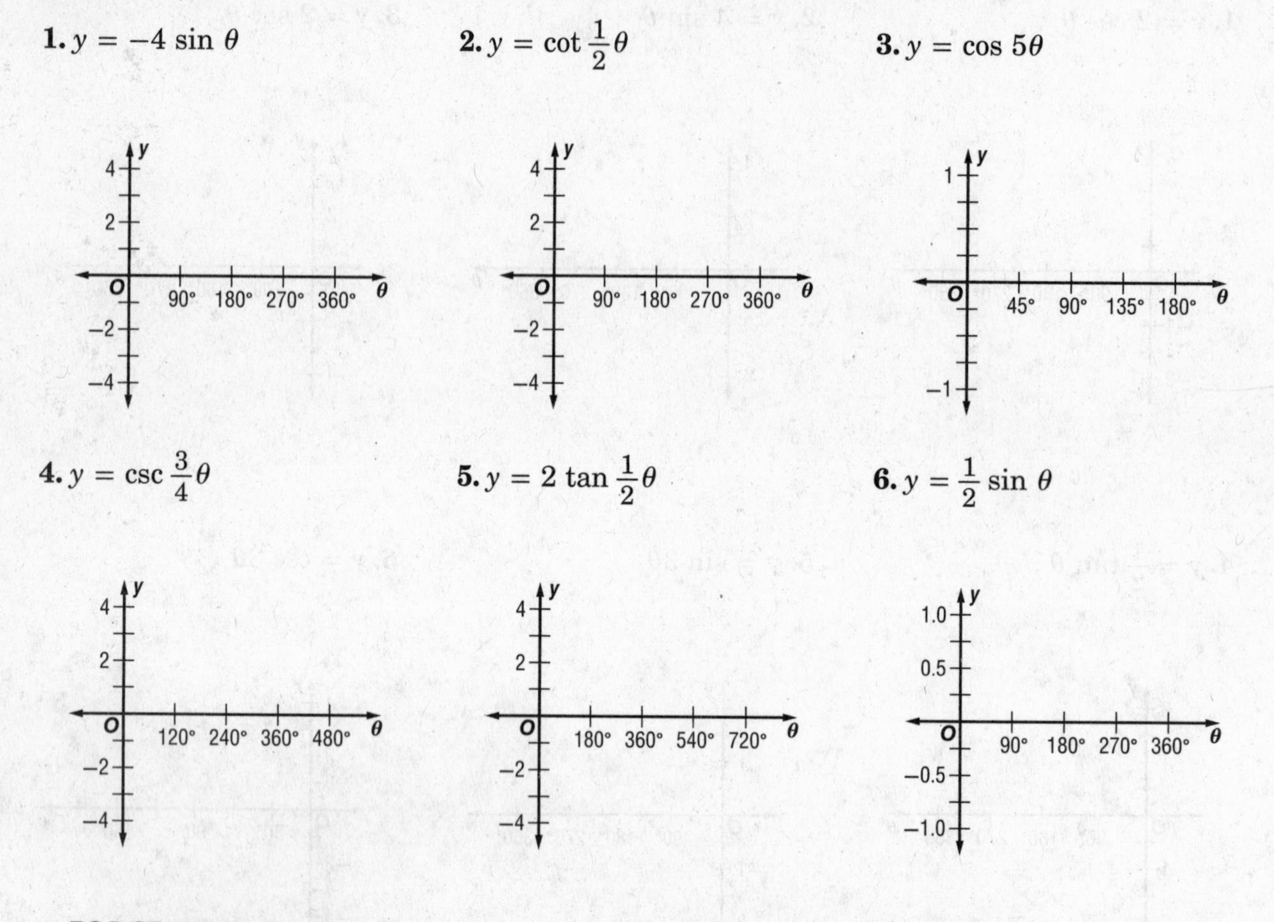

7. FORCE An anchoring cable exerts a force of 500 Newtons on a pole. The force has the horizontal and vertical components F_x and F_y. (A force of one Newton (N), is the force that gives an acceleration of 1 m/sec^2 to a mass of 1 kg.)

500 N

F_y

θ

F_x

a. The function $F_x = 500 \cos \theta$ describes the relationship between the angle θ and the horizontal force. What are the amplitude and period of this function?

b. The function $F_y = 500 \sin \theta$ describes the relationship between the angle θ and the vertical force. What are the amplitude and period of this function?

8. WEATHER The function $y = 60 + 25 \sin \frac{\pi}{6}t$, where t is in months and $t = 0$ corresponds to April 15, models the average high temperature in degrees Fahrenheit in Centerville.

a. Determine the period of this function. What does this period represent?

b. What is the maximum high temperature and when does this occur?

NAME ______________________ DATE ____________ PERIOD ____________

13-8 Skills Practice

Translations of Trigonometric Graphs

State the amplitude, period, and phase shift for each function. Then graph the function.

1. $y = \sin(\theta + 90°)$

2. $y = \cos(\theta - 45°)$

3. $y = \tan\left(\theta - \frac{\pi}{2}\right)$

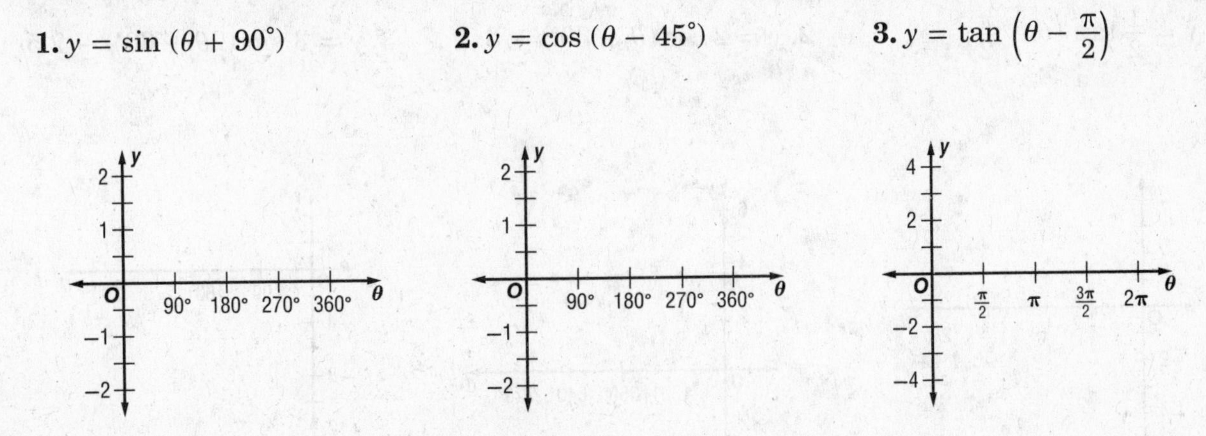

State the amplitude, period, vertical shift, and equation of the midline for each function. Then graph the function.

4. $y = \csc\theta - 2$

5. $y = \cos\theta + 1$

6. $y = \sec\theta + 3$

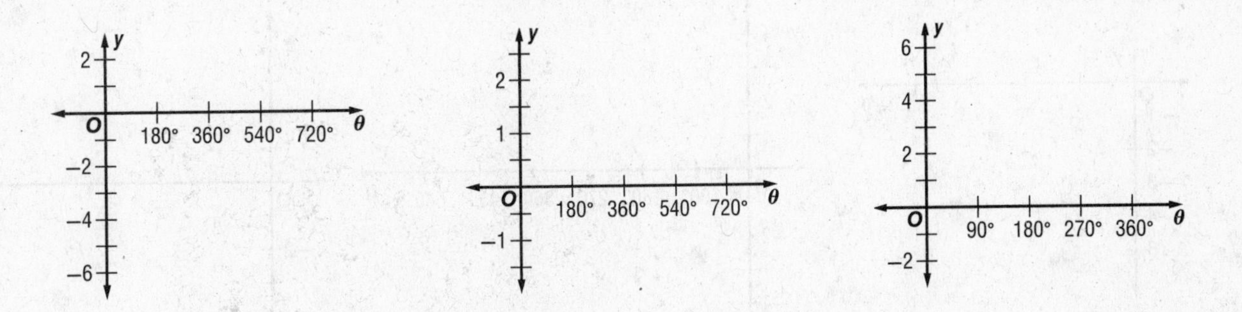

State the amplitude, period, phase shift, and vertical shift of each function. Then graph the function.

7. $y = 2\cos[3(\theta + 45°)] + 2$

8. $y = 3\sin[2(\theta - 90°)] + 2$

9. $y = 4\cot\left[\frac{4}{3}\left(\theta + \frac{\pi}{4}\right)\right] - 2$

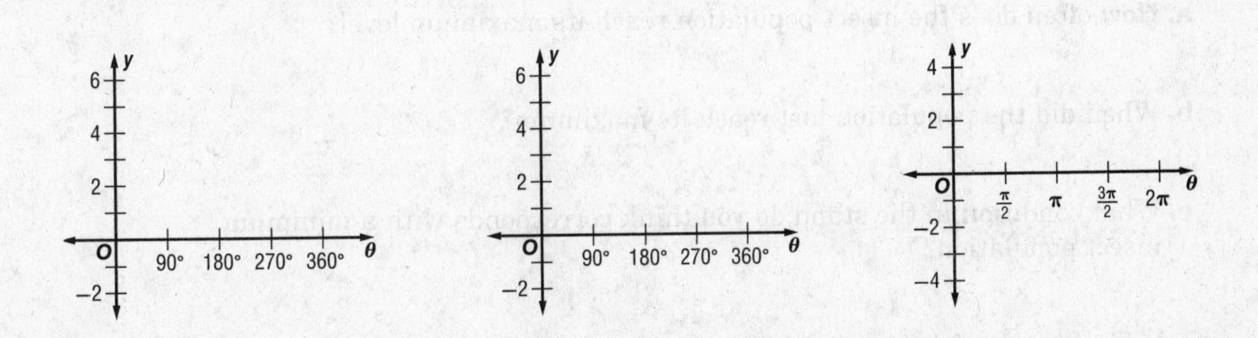

NAME ______________________ DATE ____________ PERIOD ________

13-8 Practice

Translations of Trigonometric Graphs

State the amplitude, period, phase shift, and vertical shift for each function. Then graph the function.

1. $y = \frac{1}{2}\tan\left(\theta - \frac{\pi}{2}\right)$

2. $y = 2\cos(\theta + 30°) + 3$

3. $y = 3\sin(2\theta + 60°) - 2.5$

4. $y = -3 + 2\sin 2\left(\theta + \frac{\pi}{4}\right)$

5. $y = 3\cos 2(\theta + 45°) + 1$

6. $y = -1 + 4\tan(\theta + \pi)$

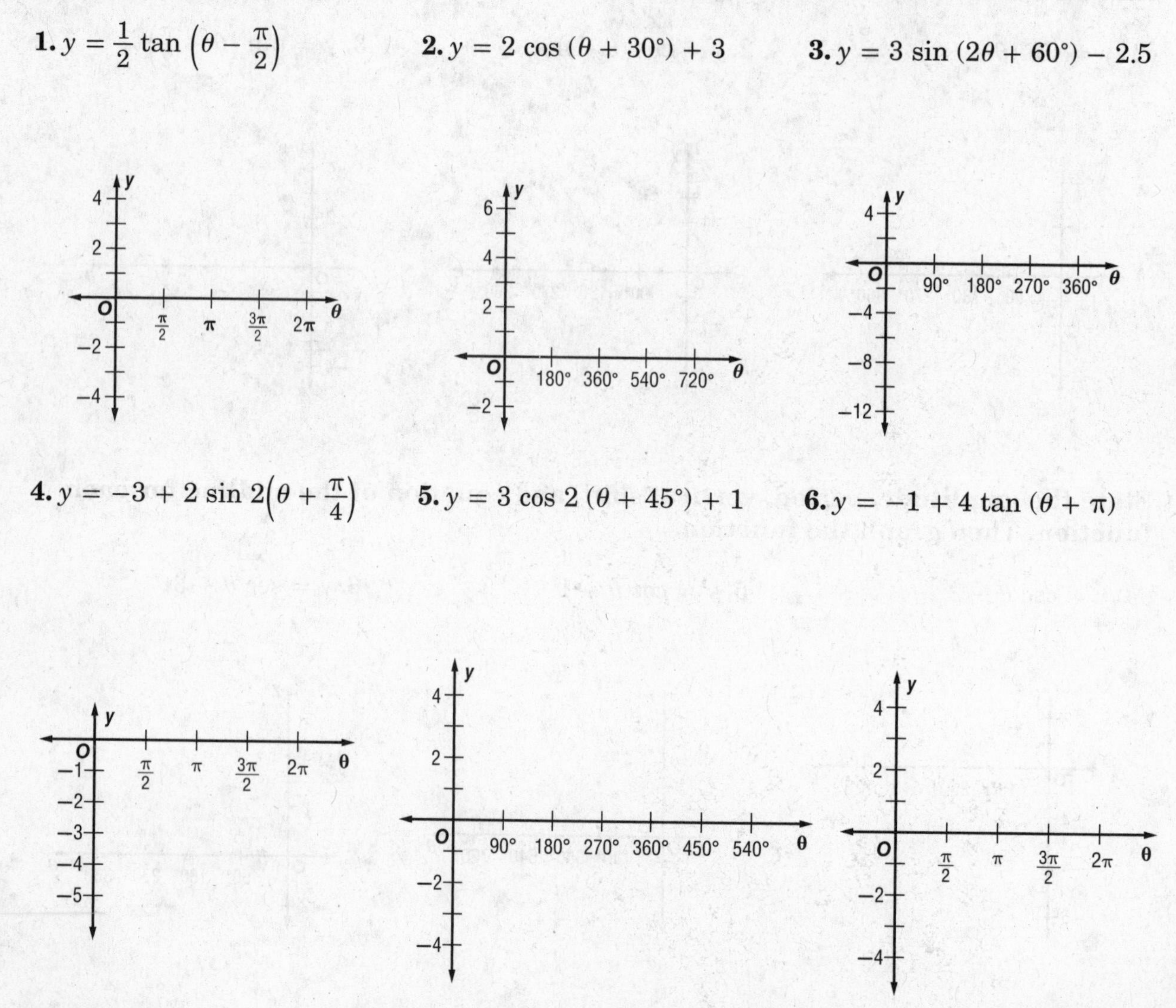

7. ECOLOGY The population of an insect species in a stand of trees follows the growth cycle of a particular tree species. The insect population can be modeled by the function $y = 40 + 30\sin 6t$, where t is the number of years since the stand was first cut in November, 1920.

a. How often does the insect population reach its maximum level?

b. When did the population last reach its maximum?

c. What condition in the stand do you think corresponds with a minimum insect population?

NAME ______________________ DATE __________ PERIOD __________

13-9 Skills Practice

Inverse Trigonometric Functions

Find each value. Write angle measures in degrees and radians.

1. $\text{Sin}^{-1} \frac{\sqrt{2}}{2}$

2. $\text{Cos}^{-1} \left(-\frac{\sqrt{3}}{2}\right)$

3. $\text{Tan}^{-1} \sqrt{3}$

4. $\text{Arctan} \left(-\frac{\sqrt{3}}{3}\right)$

5. $\text{Arccos} \left(-\frac{\sqrt{2}}{2}\right)$

6. Arcsin 1

Find each value. Round to the nearest hundredth of necessary.

7. $\sin (\text{Cos}^{-1} 1)$

8. $\sin \left(\text{Sin}^{-1} \frac{1}{2}\right)$

9. $\tan \left(\text{Arcsin} \frac{\sqrt{3}}{2}\right)$

10. $\cos (\text{Tan}^{-1} 3)$

11. $\sin [\text{Arctan} (-1)]$

12. $\sin \left[\text{Arccos} \left(-\frac{\sqrt{2}}{2}\right)\right]$

Solve each equation. Round to the nearest tenth if necessary.

13. $\cos \theta = 0.25$

14. $\sin \theta = -0.57$

15. $\tan \theta = 5$

16. $\cos \theta = 0.11$

17. $\sin \theta = 0.9$

18. $\tan \theta = -11.35$

19. $\sin \theta = 1$

20. $\tan \theta = -0.01$

21. $\cos \theta = -0.36$

22. $\tan \theta = -16.6$

NAME ______________________ DATE __________ PERIOD __________

13-9 Practice

Inverse Trigonometric Functions

Find each value. Write angle measures in degrees and radians.

1. Arcsin 1

2. $\text{Cos}^{-1}\left(\frac{-\sqrt{2}}{2}\right)$

3. $\text{Tan}^{-1}\left(\frac{-\sqrt{3}}{3}\right)$

4. $\text{Arccos}\ \frac{\sqrt{2}}{2}$

5. $\text{Arctan}\ (-\sqrt{3})$

6. $\text{Sin}^{-1}\left(-\frac{1}{2}\right)$

Find each value. Round to the nearest hundredth if necessary.

7. $\tan\left(\text{Cos}^{-1}\ \frac{1}{2}\right)$

8. $\cos\left[\text{Sin}^{-1}\left(-\frac{3}{5}\right)\right]$

9. $\cos\ [\text{Arctan}\ (-1)]$

10. $\tan\left(\text{Sin}^{-1}\ \frac{12}{13}\right)$

11. $\sin\left(\text{Arctan}\ \frac{\sqrt{3}}{3}\right)$

12. $\cos\left(\text{Arctan}\ \frac{3}{4}\right)$

Solve each equation. Round to the nearest tenth if necessary.

13. $\text{Tan}\ \theta = 10$

14. $\text{Sin}\ \theta = 0.7$

15. $\text{Sin}\ \theta = -0.5$

16. $\text{Cos}\ \theta = 0.05$

17. $\text{Tan}\ \theta = 0.22$

18. $\text{Sin}\ \theta = -0.03$

19. PULLEYS The equation $\cos \theta = 0.95$ describes the angle through which pulley A moves, and $\cos \theta = 0.17$ describes the angle through which pulley B moves. Which pulley moves through a greater angle?

20. FLYWHEELS The equation $\text{Tan}\ \theta = 1$ describes the counterclockwise angle through which a flywheel rotates in 1 millisecond. Through how many degrees has the flywheel rotated after 25 milliseconds?

14-1 Skills Practice

Trigonometric Identities

Find the exact value of each expression if $0° < \theta < 90°$.

1. If $\tan \theta = 1$, find $\sec \theta$.

2. If $\tan \theta = \frac{1}{2}$, find $\cos \theta$.

3. If $\sec \theta = 2$, find $\cos \theta$.

4. If $\cos \theta = \frac{8}{17}$, find $\csc \theta$.

Find the exact value of each expression if $90° < \theta < 180°$.

5. If $\cos \theta = -\frac{4}{5}$, find $\sin \theta$.

6. If $\cot \theta = -\frac{3}{2}$, find $\cos \theta$.

Find the exact value of each expression if $180° < \theta < 270°$.

7. If $\tan \theta = 1$, find $\cos \theta$.

8. If $\sin \theta = -\frac{\sqrt{2}}{2}$, find $\tan \theta$.

9. If $\csc \theta = -2$, find $\cos \theta$.

10. If $\cos \theta = -\frac{2\sqrt{5}}{5}$, find $\tan \theta$.

11. If $\csc \theta = -2$, find $\cot \theta$.

12. If $\sin \theta = -\frac{5}{13}$, find $\tan \theta$.

Simplify each expression.

13. $\sin \theta \sec \theta$

14. $\csc \theta \sin \theta$

15. $\cot \theta \sec \theta$

16. $\frac{\cos \theta}{\sec \theta}$

17. $\tan \theta + \cot \theta$

18. $\csc \theta \tan \theta - \tan \theta \sin \theta$

19. $\frac{1 - \sin^2 \theta}{\sin \theta + 1}$

20. $\csc \theta + \cot \theta$

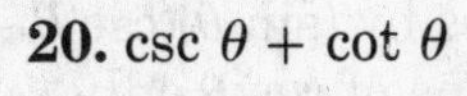

21. $\frac{\sin^2 \theta + \cos^2 \theta}{1 - \cos^2 \theta}$

22. $1 + \frac{\tan^2 \theta}{1 + \sec \theta}$

NAME ______________________ DATE __________ PERIOD __________

14-1 Practice

Trigonometric Identities

Find the exact value of each expression if $0° < \theta < 90°$.

1. If $\cos \theta = \frac{5}{13}$, find $\sin \theta$.

2. If $\cot \theta = \frac{1}{2}$, find $\sin \theta$.

3. If $\tan \theta = 4$, find $\sec \theta$.

4. If $\tan \theta = \frac{2}{5}$, find $\cot \theta$.

Find the exact value of each expression if $180° < \theta < 270°$.

5. If $\sin \theta = -\frac{15}{17}$, find $\sec \theta$.

6. If $\csc \theta = -\frac{3}{2}$, find $\cot \theta$.

Find the exact value of each expression if $270° < \theta < 360°$.

7. If $\cos \theta = \frac{3}{10}$, find $\cot \theta$.

8. If $\csc \theta = -8$, find $\sec \theta$.

9. If $\tan \theta = -\frac{1}{2}$, find $\sin \theta$.

10. If $\cos \theta = \frac{1}{3}$, find $\cot \theta$.

Simplify each expression.

11. $\csc \theta \tan \theta$

12. $\frac{\sin^2 \theta}{\tan^2 \theta}$

13. $\sin^2 \theta \cot^2 \theta$

14. $\cot^2 \theta + 1$

15. $\frac{\csc^2 \theta - \cot^2 \theta}{1 - \cos^2 \theta}$

16. $\frac{\csc \theta - \sin \theta}{\cos \theta}$

17. $\sin \theta + \cos \theta \cot \theta$

18. $\frac{\cos \theta}{1 - \sin \theta} - \frac{\cos \theta}{1 + \sin \theta}$

19. $\sec^2 \theta \cos^2 \theta + \tan^2 \theta$

20. AERIAL PHOTOGRAPHY The illustration shows a plane taking an aerial photograph of point *A*. Because the point is directly below the plane, there is no distortion in the image. For any point *B* not directly below the plane, however, the increase in distance creates distortion in the photograph. This is because as the distance from the camera to the point being photographed increases, the exposure of the film reduces by $(\sin \theta)(\csc \theta - \sin \theta)$. Express $(\sin \theta)(\csc \theta - \sin \theta)$ in terms of $\cos \theta$ only.

θ
A
B

21. WAVES The equation $y = a \sin \theta t$ represents the height of the waves passing a buoy at a time t in seconds. Express a in terms of $\csc \theta t$.

14-2 Skills Pratice

Verifying Trigonometric Identities

Verify that each equation is an identity.

1. $\tan\theta\cos\theta = \sin\theta$

2. $\cot\theta\tan\theta = 1$

3. $\csc\theta\cos\theta = \cot\theta$

4. $\dfrac{1 - \sin^2\theta}{\cos\theta} = \cos\theta$

5. $(\tan\theta)(1 - \sin^2\theta) = \sin\theta\cos\theta$

6. $\dfrac{\csc\theta}{\sec\theta} = \cot\theta$

7. $\dfrac{\sin^2\theta}{1 - \sin^2\theta} = \tan^2\theta$

8. $\dfrac{\cos^2\theta}{1 - \sin\theta} = 1 + \sin\theta$

4-2 Practice

Verifying Trigonometric Identities

Verify that each equation is an identity.

1. $\dfrac{\sin^2 \theta + \cos^2 \theta}{\cos^2 \theta} = \sec^2 \theta$

2. $\dfrac{\cos^2 \theta}{1 - \sin^2 \theta} = 1$

3. $(1 + \sin \theta)(1 - \sin \theta) = \cos^2 \theta$

4. $\tan^4 \theta + 2 \tan^2 \theta + 1 = \sec^4 \theta$

5. $\cos^2 \theta \cot^2 \theta = \cot^2 \theta - \cos^2 \theta$

6. $(\sin^2 \theta)(\csc^2 \theta + \sec^2 \theta) = \sec^2 \theta$

7. PROJECTILES The square of the initial velocity of an object launched from the ground is $v^2 = \dfrac{2gh}{\sin^2 \theta}$, where θ is the angle between the ground and the initial path h is the maximum height reached, and g is the acceleration due to gravity. Verify the identity $\dfrac{2gh}{\sin^2 \theta} = \dfrac{2gh \sec^2 \theta}{\sec^2 \theta - 1}$.

8. LIGHT The intensity of a light source measured in candles is given by $I = ER^2 \sec \theta$, where E is the illuminance in foot candles on a surface, R is the distance in feet from the light source, and θ is the angle between the light beam and a line perpendicular to the surface. Verify the identity $ER^2(1 + \tan^2 \theta) \cos \theta = ER^2 \sec \theta$.

NAME ______________________ DATE ____________ PERIOD ____________

14-3 Skills Practice

Sum and Difference of Angles Identities

Find the exact value of each expression.

1. $\sin 330°$

2. $\cos(-165°)$

3. $\sin(-225°)$

4. $\cos 135°$

5. $\sin(-45)°$

6. $\cos 210°$

7. $\cos(-135°)$

8. $\sin 75°$

9. $\sin(-195°)$

Verify that each equation is an identity.

10. $\sin(90° + \theta) = \cos\theta$

11. $\sin(180° + \theta) = -\sin\theta$

12. $\cos(270° - \theta) = -\sin\theta$

13. $\cos(\theta - 90°) = \sin\theta$

14. $\sin\left(\theta - \frac{\pi}{2}\right) = -\cos\theta$

15. $\cos(\pi + \theta) = -\cos\theta$

14-3 Practice

Sum and Difference of Angles Identities

Find the exact value of each expression.

1. $\cos 75°$

2. $\cos 375°$

3. $\sin (-165°)$

4. $\sin (-105°)$

5. $\sin 150°$

6. $\cos 240°$

7. $\sin 225°$

8. $\sin (-75°)$

9. $\sin 195°$

Verify that each equation is an identity.

10. $\cos (180° - \theta) = -\cos \theta$

11. $\sin (360° + \theta) = \sin \theta$

12. $\sin (45° + \theta) - \sin (45° - \theta) = \sqrt{2} \sin \theta$

13. $\cos \left(x - \frac{\pi}{6}\right) + \sin \left(x - \frac{\pi}{3}\right) = \sin x$

14. SOLAR ENERGY On March 21, the maximum amount of solar energy that falls on a square foot of ground at a certain location is given by $E \sin (90° - \phi)$, where ϕ is the latitude of the location and E is a constant. Use the difference of angles formula to find the amount of solar energy, in terms of $\cos \phi$, for a location that has a latitude of ϕ.

15. ELECTRICITY In a certain circuit carrying alternating current, the formula $c = 2 \sin (120t)$ can be used to find the current c in amperes after t seconds.

a. Rewrite the formula using the sum of two angles.

b. Use the sum of angles formula to find the exact current at $t = 1$ second.

NAME ______________________ DATE ____________ PERIOD ____________

14-4 Skills Practice

Double-Angle and Half-Angle Identities

Find the exact values of sin 2θ, cos 2θ, sin $\frac{\theta}{2}$, and cos $\frac{\theta}{2}$.

1. $\cos \theta = \frac{7}{25}$, $0° < \theta < 90°$

2. $\sin \theta = -\frac{4}{5}$, $180° < \theta < 270°$

3. $\sin \theta = \frac{40}{41}$, $90° < \theta < 180°$

4. $\cos \theta = \frac{3}{7}$, $270° < \theta < 360°$

5. $\cos \theta = -\frac{3}{5}$, $90° < \theta < 180°$

6. $\sin \theta = \frac{5}{13}$, $0° < \theta < 90°$

Find the exact value of each expression.

7. $\cos 22.5°$

8. $\sin 165°$

9. $\cos 105°$

10. $\sin \frac{\pi}{8}$

11. $\sin \frac{15\pi}{8}$

12. $\cos 75°$

Verify that each equation is an identity.

13. $\sin 2\theta = \frac{2 \tan \theta}{1 + \tan^2 \theta}$

14. $\tan \theta + \cot \theta = 2 \csc 2\theta$

4 Practice

Double-Angle and Half-Angle Identities

Find the exact values of sin 2θ, cos 2θ, sin $\frac{\theta}{2}$, and cos $\frac{\theta}{2}$ for each of the following.

1. $\cos \theta = \frac{5}{13}$, $0° < \theta < 90°$

2. $\sin \theta = \frac{8}{17}$, $90° < \theta < 180°$

3. $\cos \theta = \frac{1}{4}$, $270° < \theta < 360°$

4. $\sin \theta = -\frac{2}{3}$, $180° < \theta < 270°$

Find the exact value of each expression.

5. tan 105°

6. tan 15°

7. cos 67.5°

8. $\sin\left(-\frac{\pi}{8}\right)$

Verify that each equation is an identity.

9. $\sin^2 \frac{\theta}{2} = \frac{\tan \theta - \sin \theta}{2 \tan \theta}$

10. $\sin 4\theta = 4 \cos 2\theta \sin \theta \cos \theta$

11. AERIAL PHOTOGRAPHY In aerial photography, there is a reduction in film exposure for any point X not directly below the camera. The reduction E_θ is given by $E_\theta = E_0 \cos^4 \theta$, where θ is the angle between the perpendicular line from the camera to the ground and the line from the camera to point X, and E_0 is the exposure for the point directly below the camera. Using the identity $2 \sin^2 \theta = 1 - \cos 2\theta$, verify that $E_0 \cos^4 \theta = E_0\left(\frac{1}{2} + \frac{\cos 2\theta}{2}\right)^2$.

12. IMAGING A scanner takes thermal images from altitudes of 300 to 12,000 meters. The width W of the swath covered by the image is given by $W = 2H' \tan \theta$, where H' is the height and θ is half the scanner's field of view. Verify that $\frac{2H' \sin 2\theta}{1 + \cos 2\theta} = 2H' \tan \theta$.

14-5 Skills Practice

Solving Trigonometric Equations

Solve each equation for the given interval.

1. $\sin \theta = \frac{\sqrt{2}}{2}$, $0° \leq \theta \leq 360°$

2. $2 \cos \theta = -\sqrt{3}$, $90° < \theta < 180°$

3. $\tan^2 \theta = 1$, $180° < \theta < 360°$

4. $2 \sin \theta = 1$, $0 \leq \theta < \pi$

5. $\sin^2 \theta + \sin \theta = 0$, $\pi \leq \theta < 2\pi$

6. $2 \cos^2 \theta + \cos \theta = 0$, $0 \leq \theta < \pi$

Solve each equation for all values of θ if θ is measured in radians.

7. $2 \cos^2 \theta - \cos \theta = 1$

8. $\sin^2 \theta - 2 \sin \theta + 1 = 0$

9. $\sin \theta + \sin \theta \cos \theta = 0$

10. $\sin^2 \theta = 1$

11. $4 \cos \theta = -1 + 2 \cos \theta$

12. $\tan \theta \cos \theta = \frac{1}{2}$

Solve each equation for all values of θ if θ is measured in degrees.

13. $2 \sin \theta + 1 = 0$

14. $2 \cos \theta + \sqrt{3} = 0$

15. $\sqrt{2} \sin \theta + 1 = 0$

16. $2 \cos^2 \theta = 1$

17. $4 \sin^2 \theta = 3$

18. $\cos 2\theta = -1$

Solve each equation.

19. $3 \cos^2 \theta - \sin^2 \theta = 0$

20. $\sin \theta + \sin 2\theta = 0$

21. $2 \sin^2 \theta = \sin \theta + 1$

22. $\cos \theta + \sec \theta = 2$

5 Practice

Solving Trigonometric Equations

Solve each equation for the given interval.

1. $\sin 2\theta = \cos \theta$, $90^\circ \le \theta < 180^\circ$

2. $\sqrt{2} \cos \theta = \sin 2\theta$, $0^\circ \le \theta$, 360°

3. $\cos 4\theta = \cos 2\theta$, $180^\circ \le \theta < 360^\circ$

4. $\cos \theta + \cos (90 - \theta) = 0$, $0 \le \theta < 2\pi$

5. $2 + \cos \theta = 2 \sin^2 \theta$, $\pi \le \theta \le \frac{3\pi}{2}$

6. $\tan^2 \theta + \sec \theta = 1$, $\frac{\pi}{2} \le \theta < \pi$

Solve each equation for all values of θ if θ is measured in radians.

7. $\cos^2 \theta = \sin^2 \theta$

8. $\cot \theta = \cot^3 \theta$

9. $\sqrt{2} \sin^3 \theta = \sin^2 \theta$

10. $\cos^2 \theta \sin \theta = \sin \theta$

11. $2 \cos 2\theta = 1 - 2 \sin^2 \theta$

12. $\sec^2 \theta = 2$

Solve each equation for all values of θ if θ is measured in degrees.

13. $\sin^2 \theta \cos \theta = \cos \theta$

14. $\csc^2 \theta - 3 \csc \theta + 2 = 0$

15. $\frac{3}{1 + \cos \theta} = 4(1 - \cos \theta)$

16. $\sqrt{2} \cos^2 \theta = \cos^2 \theta$

Solve each equation.

17. $4 \sin^2 \theta = 3$

18. $4 \sin^2 \theta - 1 = 0$

19. $2 \sin^2 \theta - 3 \sin \theta = -1$

20. $\cos 2\theta + \sin \theta - 1 = 0$

21. WAVES Waves are causing a buoy to float in a regular pattern in the water. The vertical position of the buoy can be described by the equation $h = 2 \sin x$. Write an expression that describes the position of the buoy when its height is at its midline.

22. ELECTRICITY The electric current in a certain circuit with an alternating current can be described by the formula $i = 3 \sin 240t$, where i is the current in amperes and t is the time in seconds. Write an expression that describes the times at which there is no current.